Facing the Consequences

Facing the Consequences

Using TIMSS for a Closer Look at U.S. Mathematics and Science Education

William H. Schmidt
Michigan State University

Curtis C. McKnight
University of Oklahoma

Leland S. Cogan
Michigan State University

Pamela M. Jakwerth
American Institutes for Research

Richard T. Houang
Michigan State University

with the collaboration of:

David E. Wiley

Richard G. Wolfe

Leonard J. Bianchi

Gilbert A. Valverde

Senta A. Raizen

Christine E. DeMars

*in association with the US National Research Center for the
Third International Mathematics and Science Study (TIMSS),
Michigan State University*

KLUWER ACADEMIC PUBLISHERS
DORDRECHT / BOSTON / LONDON

A C.I.P. Catalogue record for this book is available from the Library of Congress.

ISBN 0-7923-5567-9 (HB)
ISBN 0-7923-5568-7 (PB)

Published by Kluwer Academic Publishers,
P.O. Box 17, 3300 AA Dordrecht, The Netherlands.

Sold and distributed in North, Central and South America
by Kluwer Academic Publishers,
101 Philip Drive, Norwell, MA 02061, U.S.A.

In all other countries, sold and distributed
by Kluwer Academic Publishers,
P.O. Box 322, 3300 AH Dordrecht, The Netherlands.

Printed on acid-free paper

Table of Contents

Acknowledgments

The production of this manuscript represents the culmination of an incredibly stimulating interaction among the authors with each other and the data. In addition, we are indebted to the support and insight provided by a few colleagues as we refined our thinking and developed the perspective presented in this manuscript. More specifically, we would like to express our appreciation to Shelly Naud and Elena Papanastasiou who provided technical assistance with the data, Art Kuper, April King, Sarah Kuper, and Sudip Suvedi who created initial displays, Marlene Green who provided secretarial support, and Jacqueline Babcock who supervised communications and all office functions for the US TIMSS National Research Center at Michigan State University. John Dossey, distinguished professor of mathematics at Illinois State University, Elizabeth Stage, at the New Standards project, University of California Office of the President, and Larry Suter, project director at the National Science Foundation, read an early draft and made many helpful comments for which we are grateful. Finally, the work reported here was funded through a grant from the National Science Foundation (RED-9550107). While we are grateful for the support others extended to us, the authors alone assume responsibility for the results and interpretation presented in *Facing the Consequences*.

Chapter 1
Facing the Consequences: An Overview

Today we and our children are faced with the consequences of what US science and mathematics education have become. American education is complex and has evolved over our national history. Generations of dedicated Americans – officials, school administrators, teachers, parents, and citizens– have made each educational decision they faced as carefully as possible. Those choices expressed our society's values and our citizens' beliefs about education and what is important to it. What we face now is the product of those choices. More formally, we may say that the current state of US science and mathematics education is a consequence of ingrained beliefs and practices, and an accretion of shorter-term choices made at many points within the system. Those choices reflected the deeply held beliefs and values of the citizens who made them.

If we find what we face disappointing, we must seek deeper understanding of the practices that define the education provided to American children in mathematics and the sciences. Armed with understanding we must turn to new decisions. Faced with the consequences of America's educational past, we must face them directly and choose wisely to shape a better educational future.

Most will indeed find what we face disappointing. There is certainly ample evidence to support claims that mathematics and science education in US schools are in trouble. We published some of that evidence over ten years ago.[1] We recently reported even more of that evidence.[2] This current report presents still more.

Can we any longer afford to not act on the evidence of the comparative weaknesses of US science and mathematics education? Can we ignore the data we have on the nature of those weaknesses? While our data establish only concomitance, not cause and effect, comparing mathematics and science education as practiced in many countries makes it clear that US practice is but one approach among many. Our US approach is not an inevitable consequence of the need to teach children about mathematics and the sciences. Other countries approach these same goals quite differently. Our approach is a consequence, not of any single choice or belief, but of the many values, choices, and decisions that have shaped it until now.

If the state of US science and mathematics education was unquestionably satisfactory, the complexity of choices that shape it would not matter. If it is not, the fact that education is shaped as consequences of beliefs and choices becomes crucial. Any attempt at reform, at deliberate change, must move not a single official, or group of officials, but the greater part of an entire society. The evidence we present here suggests that US mathematics and science education are *not* satisfactory. If so, the great enterprise of affecting educational choices must be undertaken. American society must be moved to take our current situation seriously and demand change. This report seeks to tell a story about our educational practice. Less directly, it seems to be a story about American educational beliefs, choices, and their consequences. If we take seriously the evidence presented regarding US mathematics and science education, then this will also be a book about facing consequences and moving forward.

TIMSS, THE THIRD INTERNATIONAL MATHEMATICS AND SCIENCE STUDY

The major evidence around which this story is built comes from data gathered as part of the Third International Mathematics and Science Study (TIMSS). TIMSS is the most extensive and far reaching cross-national comparative study of mathematics and science education ever attempted. It includes comparing the official curricula, textbooks, teacher practices, and student achievements of many countries (20 to 50 countries, depending on the particular comparison). Thousands of official documents and textbooks were analyzed. Thousands of teachers, principals, and other experts responded to survey questionnaires. Hundreds of thousands of children in almost 50 countries were tested in mathematics and science. These tests were conducted for nine-year-olds, thirteen-year-olds, and for students in the last year of secondary school.[3] The focus of this book is on a deeper set of results for nine- and thirteen-year-olds (third, fourth, seventh, and eighth grades in the US and most TIMSS countries). Only recently end-of-secondary (in the US, twelfth grade) results have been released and these are summarized briefly in Chapter 10. What follows now is a brief summary of the results released thus far that serve as a focus on the nine- and thirteen-year-olds and which are the springboard for this report.

Eighth Grade Achievement. Where did we stand compared to other TIMSS countries? The TIMSS eighth grade science results reveal the US as slightly above the international average. Students in several countries performed significantly better than those in the US (which and how many countries depends on whether one looks at the overall scores or those for a few more specific reporting categories). Many countries had scores that did not differ significantly from those of the US and several had scores significantly lower. These data were not particularly good news since we profess a desire for a high standing in science achievement among the community of nations.

The TIMSS US eighth grade mathematics results do not even paint even as pleasant a picture as the science results. Essentially, the US scored below the international mean in almost every area. US students were significantly better than only seven countries – Columbia, Kuwait, South Africa, Iran, Portugal, Cyprus, and Lithuania. There were always a large number of countries that scored significantly better than the US (although the countries and their numbers varied by whether the overall mathematics scores were compared or those for a few more specific reporting categories).

Fourth Grade Achievement. The US' brightest spot in the TIMSS achievement results was fourth grade science. The US was tied for second among TIMSS countries on the overall science score. Our students were outperformed only by those from Korea and performed significantly better than most others. While our place varied slightly among the more specific science categories, our standing was always close to the top. The same cannot be said of our fourth grade mathematics achievement results. Overall, we placed somewhat above the international average. Our rank varied among the more specific mathematics categories but did not reach the top tier of countries. US students consistently performed better in science than in mathematics.

The comparison of fourth to eighth grade for both science and mathematics shows that we consistently performed better at fourth grade than at eighth, especially in science. US students did not start behind, they fell behind. This was a simultaneous sample at two grades rather than a longitudinal study tracking the same students across grades. If there were no significant differences between the fourth grade students tested and the eighth grade students tested when they were fourth graders, and no significant differences in the aggregate curricula studied by these students, it seems fair to conclude that, were we to track the same students over the grades, we would see them fall more and more behind cumulatively. This cumulative lagging of US science and mathematics students does not bode well for the TIMSS achievement results for twelfth grade (see Chapter 10 for a snapshot of the twelfth grade results).

There is collateral evidence that US students are not inherently inferior to the students in the other TIMSS countries. Several sub-national replications of TIMSS have been done within the US. For example, a consortium (the 'First in the World' consortium) of several Illinois school districts replicated the TIMSS study and performed significantly better than the US average.[4] In fact, they performed at a level that would have placed them near the top among TIMSS countries. Further, the state of Minnesota replicated TIMSS in a sample large enough to allow characterization of the state's education as a whole and to compare this with national education systems in TIMSS countries. While its mathematics scores were only slightly better than the US average scores at eighth grade, its science scores were significantly better, comparable to those of the top performing TIMSS countries. In fact at eighth grade, in earth science, they recorded the highest score tying them with Singapore, the overall top-scoring TIMSS country.[5] While this evidence is far from systematic, it appears to demonstrate that US students, even with fairly large and representative samples, are capable of performance that would put them among the top TIMSS countries.

US Official Curricula and Textbooks. In January, 1997, the TIMSS report *A Splintered Vision* documented that US mathematics and science education reflected a fragmented system.[6] It suggested that the disarray stemmed from the lack of broad, intellectually coherent, commonly accepted, guiding visions. It pointed out that there was no single educational system in the United States, but rather many systems at the local and state levels. Each system set its own educational goals and policies. Each shaped its own state or district curricula which reflected those goals. Goals across systems shared only the broadest features. The report examined 'composite' US science and mathematics curricula built around the common features shared by local and district curricula. Those composites consisted of brief time allocated to far more topics than was typical among TIMSS countries. The goals and emphases in science and mathematics curricula was found to be strongly fragmented into many small pieces of topics common to most curricula.

Our science and mathematics curricula tended to cover far more topics in most grades than did curricula in other TIMSS countries. Furthermore, topics tended to remain in US curricula for more grades than was typical in the other countries. As a result, most official mathematics or science curricula required covering many topics in each grade. The operative word is 'covering.' Since instructional time for mathematics and the sciences was limited, and did not vary greatly among TIMSS countries, the time allocated to each topic was necessarily small in countries such as the US that covered many topics. This led *A Splintered Vision* to summarize US mathematics and science curricula as being 'a mile wide and an inch deep.'

US textbooks were similarly characterized. To be commercially viable, adopted by states and selected by districts, textbooks have had to include materials that reflect the mathematics and science curricula of many state and local educational systems and to support teachers in providing instruction on required topics. *A Splintered Vision* suggested that, unfortunately, in the US context of 'many curricula, many goals,' a rational strategy for commercial textbook publishers was to make their textbooks as broad as possible. A similar argument could be made for commercial 'standardized' tests and even some state assessments.

As a result, US textbooks and tests reflected our curricula's fragmentation by covering almost every topic at each grade level and, in the process, becoming physically the largest and heaviest textbooks among all TIMSS countries. Since they became all-inclusive, they seldom have taken stands on which content was most central or strategic. They offered little guidance to teachers in selecting from their inclusive contents and likely contributed to the number of topics teachers considered and the brief time given to each. They have become a part of the problem rather than a part of the solution.

US Mathematics and Science Teachers. US teachers have been directed to take official curricula seriously and teach the best they can with the resources available. The demands of schools as workplaces make them rely on their textbooks extensively. *A Splintered Vision* suggested that as a result of fragmentation in curricula and textbooks, our teachers commonly tried to accomplish a mosaic of fragmented, small tasks. Our mathematics and science classrooms reflected the fragmentation created by our many local and state educational systems' diverse goals. The teachers varied greatly in their instructional goals and practices.

One study, funded to support the US' participation in TIMSS, collected videotapes of several eighth grade mathematics classrooms included in the US, Japanese, and German samples for the TIMSS main survey. These videotapes were coded and analyzed, both by researchers for aspects of instructional strategy and by a panel of mathematicians who sought to assess the 'coherence' of the mathematics presented. The data suggested that US classes, far more than the other countries, moved among different, short duration activities and far more often allowed students to begin their 'homework' in class (findings corroborated by the TIMSS survey data on instructional practices). Mathematicians judged the mathematics presented in US classrooms to be far less coherent and less effective than that in the other two countries' classrooms, especially those of Japan.

How can we understand these findings? We now have empirical data on many aspects of US mathematics and science education and additional findings are revealed as these data are further analyzed. How can we combine these data and results to understand best what we have found? We need to understand why there were achievement differences between mathematics and science in the US, why there were differences between fourth and eighth grades, why sub-national results were superior to the US averages, how our curricula and textbooks became so fragmented, and the resulting differences in instructional practices. We believe the differences are tied to access to educational possibilities in US science and mathematics education and to differing access for differing students.

The instrument development and analysis plans for TIMSS have been driven from their beginnings by a model of educational opportunity – another way of characterizing the distribution of access to educational possibilities. This model focuses on four fundamental questions (following a suggestion by D. E. Wiley). In their simplest form, the four questions are:

1. What were students expected to learn (from educational opportunities encountered in schooling)?

2. Who delivers the instruction (by which students have opportunities to learn)?

3. How is instruction organized (in shaping and distributing opportunities to learn)?

4. What have students learned? [Considered in conjunction with these previous questions]

The questions above interact with levels of educational systems to produce a complex, detailed model of how educational opportunities are intentionally formed, articulated, shaped in various aspects of their implementations, realized in classroom events, and, indirectly, shape student outcomes through their distribution. Alternatively, the distribution of educational opportunities may be referred to as access to educational possibilities, a phrase we prefer as it emphasizes the distinction between what is made available to students (possibilities) and what students do with the activities offering possibilities (experiences). Further details of this model can be found in Schmidt and McKnight (1995)[6].

Access to educational possibilities is considered a significant construct for characterizing educational systems and children's school experiences. Throughout the remainder of this volume, we will explore what we believe to be true as supported by evidence from the TIMSS data about this access and how it is differentiated for different kinds of students. We believe that these ideas go a considerable distance in helping us to understand what we observe in US science and mathematics education.

Our discussion examines education system features one at a time. However, this should not be taken to imply that we ignore interactions among those features or seek to remove them from their systemic context. Certainly it should not be taken to imply that we in any way suggest that single-factor explanations of science and mathematics achievement are adequate. Many of the things we discuss for the US are done in some high achieving countries as well. It is a temptation to conclude that, since these factors seem the same in some higher and some lower achieving countries, that they are unimportant as explanations for achievement. However, if we take seriously our position that mathematics and science education problems are systemic then we must think systemically – about how these factors interact and relate to their educational and cultural context to possibly impact achievement. We must not ask simplistically if each, as an isolated factor, is related to mean achievement.

FACING THE CONSEQUENCES

The further analysis of the TIMSS data make a more detailed picture possible and allow us to take a closer, deeper look at what lies behind this picture of the state of US mathematics and science education. This report is devoted to presenting the results of that closer look. Exhibit 1.1 shows the overall mathematics and science achievement results already released for thirteen-year-olds and indicates the countries that will serve as the focus of this report.

Exhibit 1.1. Overall Mathematics and Science Achievement for Eighth Grade* Students Compared to the US with Focus Countries for this Report Indicated in Bold.

Mathematics		Science	
Nation	**Average**	**Nation**	**Average**
Singapore	643	Singapore	607
Korea	607	Czech Republic	574
Japan	605	Japan	571
Hong Kong	588	Korea	565
Belgium (Fl)	565	Bulgaria	565
Czech Republic	564	Netherlands	560
Slovak Republic	547	Slovenia	560
Switzerland	545	Austria	558
Netherlands	541	Hungary	554
Slovenia	541	England	552
Bulgaria	540	Belgium (Fl)	550
Austria	539	Australia	545
France	538	Slovak Republic	544
Hungary	537	Russian Federation	538
Russian Federation	535	Ireland	538
Australia	530	Sweden	535
Ireland	527	United States ••••••••	534
Canada	527	Germany	531
Belgium (Fr)	526	Canada	531
Sweden	519	Norway	527
Thailand	522	New Zealand	525
Israel	522	Thailand	525
International	513	Israel	524
Germany	509	Hong Kong	522
New Zealand	508	Switzerland	522
England	506	Scotland	517
Norway	503	Spain	517
Denmark	502	*International*	516
United States ••••••••	500	France	498
Scotland	498	Greece	497
Latvia (LSS)	493	Iceland	494
Spain	487	Romania	486
Iceland	487	Latvia (LSS)	485
Greece	484	Portugal	480
Romania	482	Denmark	478
Lithuania	477	Lithuania	476
Cyprus	474	Belgium (Fr)	471
Portugal	454	Iran, Islamic Republic	470
Iran, Islamic Republic	428	Cyprus	463
Kuwait	392	Kuwait	430
Columbia	385	Columbia	411
South Africa	354	South Africa	326

☐ Significantly Higher
☐ No Significant Difference
▨ Significantly Lower

* See note 3 on page 12.

Source: Beaton, A.E., Martin, M.O., Mullis, I.V.S., Gonzalez, E.J., Smith, T.A., and Kelly, D.L. (1996). *Mathematics Achievement in the Middle School Years: IEA's Third International Mathematics and Science Study*. Chestnut Hill, MA: Center for the Study of Testing, Evaluation, and Educational Policy, Boston College, p.22.

Beaton, A.E., Martin, M.O., Mullis, I.V.S., Gonzalez, E.J., Smith, T.A., and Kelly, D.L. (1996). *Science Achievement in the Middle School Years: IEA's Third International Mathematics and Science Study*. Chestnut Hill, MA: Center for the Study of Testing, Evaluation, and Educational Policy, Boston College, p.22.

We conjecture in a later part of this report that the state of US science and mathematics education reflects the consequences of the beliefs and choices that have shaped our state, and local educational systems as well as efforts at the federal level. The data establish the 'what's' of US mathematics and science education. They do not establish cause and effect but, suggest only the 'why's' of what they find occurring in current mathematics and science education.

The TIMSS data allow a closer look at life in our schools and their science and mathematics classrooms and suggest that we are unfocused and incoherent in what we seek to accomplish and how we pursue it. That investigation permits the hypothesis that many adjustments to 'the way things are' have become an accepted part of educating our children in our state and local systems. They have become the norm for coping with our fundamental fragmentation. Because of our national tradition and a widely shared commitment to local control of education, responsibilities have been distributed among many educational systems within the country. Americans distribute educational responsibilities so consistently to states and local districts that we cannot meaningfully speak of a single US education system but only of 'educational systems.' That traditional choice, so deep as to approach being a part of an American creed, can be defended politically, ideologically, and even philosophically. We seek not to argue its merits but rather to link it to characteristic features in US mathematics and science education.

This tradition of distributed responsibility is further extended by some current reform efforts in education such as site-based management of schools, multicultural education, magnet and charter schools, vouchers, and the running of public schools by private corporations. These initiatives are based on the belief that individual principals, teachers, and parents can make better educational choices than the local or state officials or systems. Our tradition of distributed responsibility is further strengthened by the current general trend to rearrange authority and decision-making so that more resides in states and communities and less at the federal level. In this volume, we will use the term education systems (in the plural) to emphasize that, in the US, responsibility and decision making in educational matters are widely distributed.

Americans do not deliberately choose to keep their children from being their best. Adjustments often arise when we, as well-intentioned citizens, try to make the best choices we can in local educational situations which often form the horizon of our educational insight. Many of our educational systems' choices are made in relative isolation with only a picture of what happens locally and limited insight into what is happening nationally.

Examined only in a local context, these choices appear to create planned, orderly educational activities that lead to reasonable outcomes and achievements. It is hard to see how dysfunctional our system has become. We measure ourselves most often by local hopes and local successes. We look only rarely at the national mosaic of our combined efforts. We look even more rarely at our achievements and seemingly well planned efforts against international benchmarks. However, when we do so, we see how ineffectively our educational systems function in sharing mathematics and the sciences with our children. Only then do we see how limited and unsatisfactory our successes are.

We, the authors of this report, believe that US mathematics and science education currently is in an undesirable state and that the education systems producing the *status quo* must be improved. What US education systems do, we conjecture, are almost certainly consequences of society's beliefs and choices – the systems do what we, the people, expect them to do. If we as Americans are getting results that disappoint us and are not what we expected, we must not simply blame schools, or teachers, but instead must examine our own beliefs and choices. TIMSS holds a mirror to US education and its outcomes: by looking at more than US educational practice, it makes thinking outside the box of long-established US beliefs and practices possible. It allows us to reflect on the ways choices are made and the resulting consequences for students.

A 'SITE MAP' OF THIS BOOK

This is a report in several parts. The first part looks at TIMSS data relevant to practices we follow in mathematics and science education as compared to practices in other TIMSS countries. The second part takes a closer look at US science and mathematics achievement to understand more clearly how this is related to educational practice. The third part, frankly more speculative (especially Chapter 7), offers one hypothesis we feel is likely implied by the data presented in Chapters 2 through 6. Other chapters in this part go on to summarize some of the lessons we believe have been learned through this closer look at the TIMSS data.

Part I. The first part of this report, *Choices, Beliefs, Consequences,* – about what we teach, to whom we teach it, and how we organize that instruction – is dedicated to exploring characteristic features of our educational practices.

How has the US shaped mathematics and science education? The answer lies in some of the more basic facts about classroom instruction: how much time teachers spend on mathematics or science, what content they teach, to whom they teach it, and how they teach it. One can even make a case for when – at what grade level – the content is taught. Our many education systems decide the who, what, how, how long, and when for teaching mathematics and science at various grade levels.

Chapter 2, *What We Teach, Who We Teach,* surveys in more detail our choices for what we teach in the sciences and mathematics and to whom we teach those things. It examines the time allocated to mathematics and science instruction and to their various topics. It also examines the topics that are covered and emphasized both from the perspective of the 'official' as it resides in curricula and mandated textbooks, and of the 'actual', as it resides in what our teachers teach. It also examines to whom access to various contents is given.

Chapter 3, *How We Teach*, continues by examining how we teach mathematics and the sciences in US schools. It examines what teachers say they have done in instruction and what students say they have experienced.

Chapter 4, *Schools, Teachers, Students, and Other Factors*, concludes by examining other characteristics. These include aspects of our formal system such as staffing patterns and school characteristics, as well as characteristics of student interest and involvement.

Part II. We used TIMSS in Part I to look at what is taught in mathematics and the sciences, to whom it is taught, and how it is taught. In Part II, *Our Students' Accomplishments*, we turn to the likely correlates of America's traditional educational practices for our students and their accomplishments in mathematics and science. We examine US students' accomplishments as they emerged in the TIMSS achievement results thus far released. We move beyond global, highly aggregated reports of achievement comparisons to study these results in more detail. We seek to examine the results in ways that are more sensitive to the effects of curricula in mathematics and science, so that if curricula really do matter to achievement, we will better understand how this works.

In Chapter 5, *Curriculum Does Matter*, we discuss the fact that, contrary to appearances in highly aggregated, scaled test results, achievement *does* relate to mathematics and science curricular differences. This should seem obvious. If achievement is not related to curricula and what happens in school, why gather achievement data at all? Surely it is not for the purpose of declaring winners and losers. The US and most other countries participate in cross-national studies such as TIMSS to learn more about what is working well and what is not working well in their education systems. If achievement scores cannot be related to changeable educational factors (that is, cannot be related to making better educational choices), there seems little point in going through the massive efforts needed to gather cross-national information. This does not have to be so, as we try to demonstrate by more detailed analyses reported in this section.

In Chapter 6, *Access to Curriculum Matters*, we go further. The data suggest more than the fact that achievement differences are systematically related to curricula and other systemic factors. The data suggest that these differences are also related to students having access to different possibilities for learning mathematics and science. Not surprisingly, access seems strongly related to differing achievement. This not only further shows that curriculum matters, but calls into serious question policies leading to such differences. We think the data show that the curricular differences related to access to educational possibilities do matter. They affect overall student achievement. They also influence how much students can achieve even by working hard and to the best of their abilities.

Part III. In Parts I and II, we present the details of various stories told through data – empirical findings from TIMSS. We look at what mathematics and science were taught in US schools, how they were taught, who was taught what content, how schools functioned, what US teachers and students believed, and how curriculum and differing access to curriculum mattered for educational achievements and gains. In Part III, *Is There an Underlying Story?* we turn to another sort of analysis to attempt an answer to the question: Is there a main story underlying the details of all these empirical portraits of US science and mathematics education?

In Chapter 7, *Systemic Features, Following Consequences: A Hypothesis*, we examine what we believe to be among the more fundamental consequences of US educational choices: structural characteristics of educational practice in America and its education systems that are formed by American beliefs. While this is clearly not the only possible explanation for what is observed in the TIMSS data, these systemic characteristics seem linked to educational practice in important ways. We believe them worth exploring for the light they shed on what is likely to work and not to work in efforts to change US science and mathematics education. We also summarize the results thus far and integrate them with results from previously released TIMSS reports. In this section, we discuss the more visible features of US science and mathematics education implied by the structural characteristics discussed in the immediately preceding section. In this way, we hope both to organize some of the implications of the TIMSS data and to suggest an underlying story about fundamental consequences of deeper but less visible educational choices in America.

What does it tell us about what we must do to secure a better educational future for American children learning mathematics and science? Chapter 8, *There Are No Magic Bullets*, acknowledges that there is not a single factor, even curriculum, that makes all the difference. We discuss the unsuccessful search for monolithic solutions to our systemic problems in mathematics and science education. We also discuss what this suggests about the likelihood of finding such a 'quick fix' in the future.

In Chapter 9, *Some Stories TIMSS Can and Cannot Tell*, we summarize some of the main lessons we have learned from examining more closely the currently available TIMSS results. We also discuss what kinds of supplementary research might be most helpful in the future. Finally, in Chapter 10, *What's the Next Story?* we briefly discuss the last release of TIMSS achievement data – that for end of secondary school students – and what remains to be seen of its implications in future empirical investigation.

Notes–

[1]   See McKnight et al. (1987). *The Underachieving Curriculum.* Champaign, IL: Stipes.

[2]   See: Schmidt, W. H., McKnight, C. C., and Raizen, S. A. (1997). *A Splintered Vision: An Investigation of U.S. Science and Mathematics Education.* Dordrecht/Boston/London: Kluwer; National Center for Education Statistics (NCES). (1996). *Pursuing Excellence: A Study of U.S. Eighth-Grade Mathematics and Science Teaching, Learning, Curriculum, and Achievement in International Context* (NCES 97-198). Washington D.C.: U.S. Department of Education. National Center for Education Statistics; National Center for Education Statistics (NCES). (1997). *Pursuing Excellence: A Study of U.S. Fourth-Grade Mathematics and Science Achievement in International Context* (NCES 97-255). Washington D.C.: U.S. Department of Education. National Center for Education Statistics; and National Center for Education Statistics (NCES). (1998). *Pursuing Excellence: A Study of U.S. Twelfth-Grade Mathematics and Science Achievement in International Context* (NCES 98-049). Washington D.C.: U.S. Department of Education. National Center for Education Statistics.

[3]   The official TIMSS student populations were defined as the two adjacent grades containing the majority of nine-year-olds (Population 1), the two adjacent grades containing the majority of thirteen-year-olds (Population 2), and students in the final year of secondary school (Population 3). Population Three had two major parts: a representative sample of all students in the last year of secondary school a country offered and a representative sub-sample of mathematics or science 'specialists' (those taking calculus, physics, or both, all three categories having their own sub-sample). The general end-of-secondary school population was tested on science and mathematics literacy. The specialists were given the literacy tests and also tests appropriate to their specialized sub-sample (calculus, etc.). In the US and most TIMSS countries, the upper grade of student populations one and two corresponded to the equivalent of grades four and eight respectively. Throughout this report we use 'fourth' and 'eighth grade' to refer to these students. More information about the students included in these populations in each of the countries included in this report may be found in Appendix A–Table 1.

[4]   Kroeze, D., & Johnson, D. (1997). *Achieving Excellence: A Report of Initial Findings of Eighth Grade Performance from the Third International Mathematics and Science Study, First in the World Consortium.* Oak Brook, IL: North Central Regional Educational Laboratory (NCREL).

[5]   SciMath^{MN}. (1997). *Minnesota TIMSS Report: A Preliminary Summary of Results.* St. Paul, MN: SciMath^{MN}.

[6]   Schmidt, W. H., & McKnight, C. C. (1995). Surveying educational opportunity in mathematics and science: An international perspective. *Educational Evaluation and Policy Analysis, 17,* 337-353. Discussion of the model may also be found in: Schmidt, W. H. et al (1996). *Characterizing Pedagogical Flow: An Investigation of Mathematics and Science Teaching in Six Countries.* Dordrecht/Boston/London: Kluwer; Schmidt, W. H., McKnight, C., Valverde, G. A., Houang, R. T., & Wiley, D. E. (1997). *Many Visions, Many Aims, Volume I: A Cross-National Investigation of Curricular Intentions in School Mathematics.* Dordrecht/Boston/London: Kluwer; and Schmidt, W. H., Raizen, S. A., Britton, E. D., Bianchi, L. J., & Wolfe, R. G. (1997). *Many Visions, Many Aims, Volume II: A Cross-National Investigation of Curricular Intentions in School Science.* Dordrecht/Boston/London: Kluwer.

Part I
Choices, Beliefs, Consequences

Looking at how other TIMSS countries have organized their approaches to mathematics and science education, it is clear that there is no one way or 'natural' way to create such a system. In that sense US education is a matter of choice – in truth, the combination of many choices – since clearly there are other ways in which US science and mathematics education could have been accomplished. Those choices and the beliefs that underlie them inevitably affect what is taught, who is taught, how teaching is carried out, and other aspects of mathematics and science education. The chapters in Part I, *Choices, Beliefs, Consequences*, present a closer empirical look at these important aspects of mathematics and science education. We survey what we teach in the sciences and mathematics and to whom we teach those things (Chapter 2), how we teach them (Chapter 3), and other characteristics of mathematics and science education (Chapter 4).

Part I
Choices, Beliefs, Consequences

Chapter 2
What We Teach, Who We Teach

Important characteristics of the current US educational system and practice are reflected in some simple facts about classroom instruction: how long we teach mathematics or science, what content we teach, to whom we teach the content, and how it is taught. One can even make a case for choices being reflected in when – at what grade level – the content is taught. Our many education systems (local, state, and national) decide the who, what, how, how long, and when for teaching mathematics and science at various grade levels. What are these choices like and how similar are they among the US systems?

HOW DO WE ALLOCATE OUR
MATHEMATICS AND SCIENCE TIME?

We begin with how much time is devoted to mathematics and science instruction, one of the more global features characterizing instructional practice. Do schools within the US vary markedly in how much classroom time they devote to mathematics and science? If there are differences, are they more at some grade levels than others? Are they more for mathematics or for science? Are the differences significant when viewed in a cross-national context?

Exhibit 2.1a on page 16 shows how the number of instructional hours per year in eighth grade mathematics varied for several TIMSS countries.[1] The US clearly devoted as much or more time to mathematics instruction as most other countries. The lower quartile of US instructional time (the bottom of the box) in each case was more than 135 hours per year. While this was not higher than the extremes in all other countries, it was higher than the upper quartile of around half of the remaining countries. It was clearly higher than even the extremes in Korea, Sweden, Thailand as well as in Belgium (Flemish), Japan, the Netherlands, and Norway – all of which showed virtually no variation. This pattern held true when there were not different mathematics courses at eighth grade, and when there were as well as when instructional time in both the least and most advanced courses of study were considered. Clearly the US spent as much, if not more, instructional time on eighth grade mathematics than most other countries.

Exhibit 2.1a. *Variation in the Reported Math Instructional Hours for Eighth*
Grade* Students.

Number of instructional hours/year in math for students and the percent of students following the
least advanced math course in schools offering more than one math course.

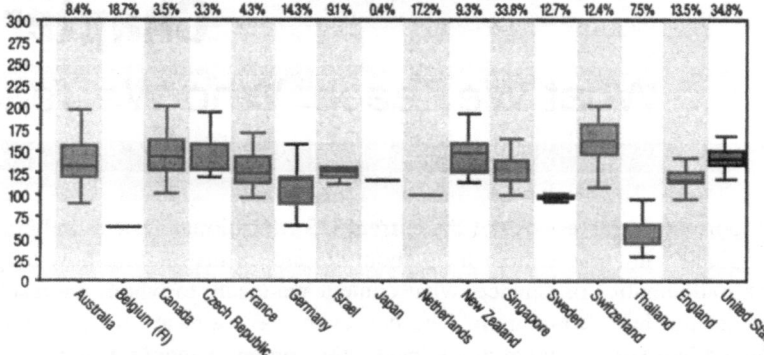

Number of instructional hours/year in math for students and the percent of students in schools
offering only one math course.

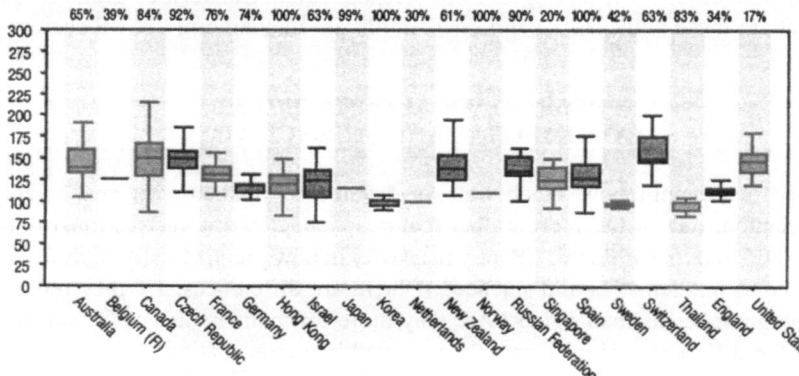

Number of instructional hours/year in math for students and the percent of students following the
most advanced math course in schools offering more than one math course.

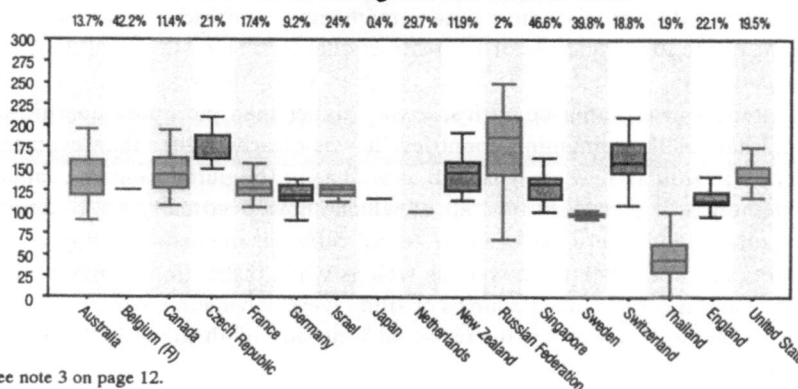

* See note 3 on page 12.

Exhibit 2.1b. *Variation in the Reported Science Instructional Hours for Eighth Grade* Students.*

Number of instructional hours/year in science for students and the percent of students following the *least advanced* science course in schools offering more than one science curricular types.

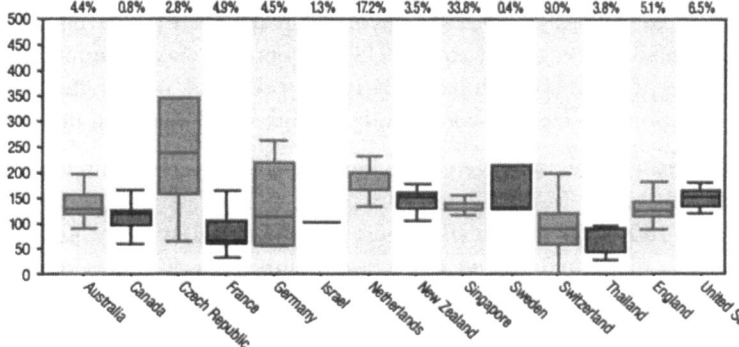

Number of instructional hours/year in science and the mean percent of students in schools offering only one science curricular type.

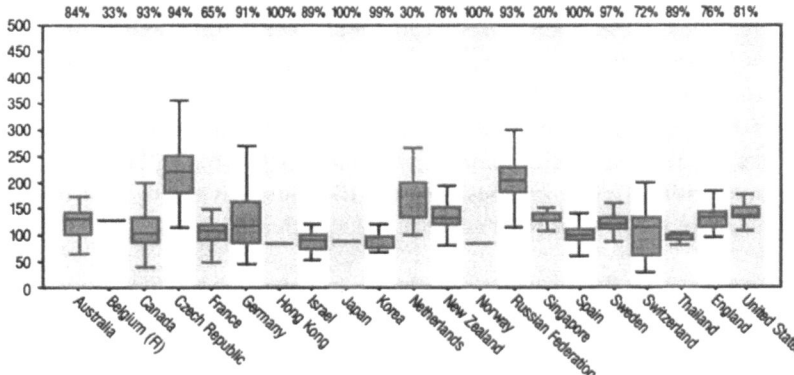

Number of instructional hours/year in science for students and the percent of students following the *most advanced* science course in schools offering more than one science curricular types.

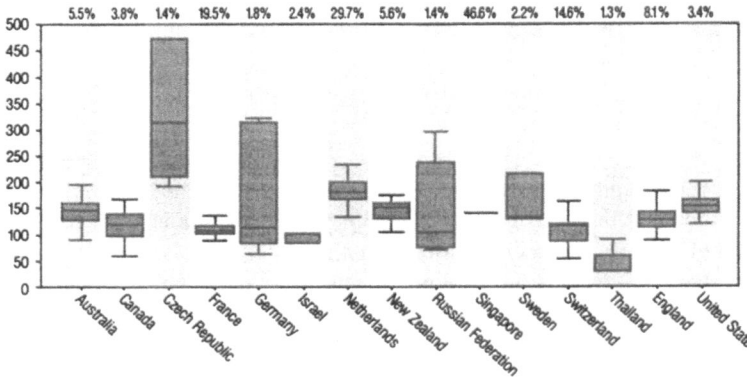

The US had greater variation in instructional time than did many of the other countries, at least when measured by the spread of instructional times for the middle 50 percent of the sample (the length of the box in Exhibit 2.1). There were some exceptions, however, most of which occurred for the least and most advanced courses of study. For example, Thailand, Canada, and the Czech Republic showed considerably more variation than the US for the middle 50 percent of classes for students involved in the least advanced course of study. Even so, the variation among US mathematics classes in total instructional hours per year was not particularly great. Further, despite the generally larger variation, the US was consistently as high, or higher, than the others.

There was little overall difference within the US between eighth grade mathematics and science instructional hours per year. The median for mathematics was about 150 hours in schools where only one eighth grade mathematics course was offered and around 141 to 147 hours when different courses were offered. This compared to a median of about 135 hours for eighth grade science. For schools which offered more than one type of eighth grade science course the median was slightly more (about 155 hours per year). The average instructional time per day was about 45 minutes and did not vary greatly.

This must not be taken to imply that there were no differences in the number of mathematics instructional hours per year within US schools. Instructional time varied by over 25 hours in schools where all students took the same eighth grade mathematics course, a difference equivalent to several weeks' instruction. The differences were slightly less but still over 20 hours in schools with different courses both for the most advanced class offered by each school and for the least advanced class offered.

Exhibit 2.2. Percents* of US Eighth Grade Students in Schools Offering Differently Titled Mathematics Courses.

Number of Different Titles	Percent of Students
Single Class School	5.4 (1.9)
One	21.0 (3.0)
Two	37.8 (3.7)
Three	29.6 (3.5)
Four	4.7 (1.5)
Five	1.5 (1.0)

* In all displays standard errors are included in parentheses.

Many US schools that taught eighth grade mathematics offered more than one course. Exhibit 2.2 shows the percentages of students in schools that offered differently-titled eighth grade mathematics classes. The exhibit makes it clear that there were significant numbers of students in US eighth grade mathematics classes that found themselves in courses that bore different titles and had substantive differences from other courses in which some of their fellow students in the same school studied mathematics. About one-fifth of the sampled students in eighth grade mathematics were in schools that offered only one course type. Just over one-third were in schools that offered two types of eighth grade mathematics (most often some form of general mathematics and some form of algebra). That left over one-third that were in schools that offered three or more courses. About 75 percent of the sampled students were in schools with at least two mathematics courses. Since placement in the different course types was unlikely to be random or by choice, these data indicate how pervasive tracking was – and likely still is – in US eighth grade mathematics.

Exhibit 2.1 suggests that the variance in instructional time was fairly consistent among course types, from the most to the least advanced types within each school that offered more than one type. The median instructional hours per year were also similar among these class types in US eighth grade mathematics.

By contrast, several other countries – for example Norway, Korea, Hong Kong, Spain, and Japan – essentially had only one course type for eighth grade mathematics (as reported by one hundred percent of the sampled schools in each of these countries). Exhibit 2.3 on page 20 shows how much 'class type' variation there was for some TIMSS countries. In fact, the majority of students attended schools in which there was only one course type, with the exception of six countries. This implies that most children in most countries attended schools that offered only one course type.

Other data collected at the country (national) level from their official curriculum guides indicates that there were common curricula within almost all of these countries. What the school data in some countries might reflect was that different sections of the same course type (thus having the same content) were designed for students with different abilities. Note that this was for a small percentage of students (less than one-fourth) in all but a few countries. Furthermore, instructional time varied little which suggests that it was set at a fixed number by a policy that was reasonably well met (or reported as having been met) by most schools and teachers in these countries.

There were even fewer cases in which several different curricular types were reported as offered in eighth grade science. Exhibit 2.4 on page 21 shows data for eighth grade science students and different curricular types which corresponds to that of Exhibit 2.3 for eighth grade mathematics. For eighth grade science, most countries portrayed had 90 to 100 percent of their students in schools which offered only one curricular type for the grade. Even the US – highly differentiated in mathematics – had over 80 percent of its students in schools in which only one curricular type of eighth grade science was taught.

Far fewer US students were in schools with two or three courses for eighth grade science (about 10 and 5 percent, respectively). Even so, the variance in instructional time in US eighth grade science classrooms (with a median of about 136 hours per year and 45 minutes per day) was about the same as that for mathematics. For those schools that offered two or more science course types (about 20 percent of the students), the distribution of instructional time was similar. Some students attended schools which allocated only about 40 minutes per day for science and others attended schools allocating about 70 minutes per day. As a result, eighth grade science instructional time for schools offering more than one course type was similar to that of mathematics, but with medians of 153 to 154 hours per year.

Exhibit 2.3. *Percent of Eighth Grade Students in Schools Offering Various Numbers of Mathematics Courses.*

	1 Course	2 Courses	3 or More Courses
USA	17.4 (3.4)	35.0 (4.3)	47.6 (4.5)
Australia	64.7 (4.0)	11.3 (2.9)	24.0 (3.7)
Belgium (Fl)	39.2 (2.6)	60.8 (2.6)	0.0
Canada	83.6 (3.0)	12.7 (2.7)	3.7 (1.6)
Czech Republic	94.6 (2.9)	5.4 (2.9)	0.0
England	35.4 (4.6)	8.4 (2.4)	56.1 (4.9)
France	100.0	0.0	0.0
Germany	74.6 (4.2)	25.4 (4.2)	0.0
Hong Kong	100.0	0.0	0.0
Israel	72.5 (9.3)	16.5 (7.8)	11.0 (6.2)
Japan	99.3 (0.7)	0.7 (0.7)	0.0
Korea	100.0	0.0	0.0
Netherlands	30.1 (5.0)	25.8 (5.5)	44.1 (6.0)
New Zealand	60.9 (4.2)	9.7 (2.5)	29.4 (4.0)
Norway	100.0	0.0	0.0
Russian Federation	90.8 (2.7)	7.8 (2.6)	1.3 (0.8)
Singapore	19.6 (3.9)	80.4 (3.9)	0.0
Spain	100.0	0.0	0.0
Sweden	42.4 (5.8)	48.4 (5.4)	9.1 (2.8)
Switzerland	63.3 (4.0)	17.5 (3.1)	19.2 (3.2)
Thailand	84.7 (4.3)	15.3 (4.3)	0.0

The situation was somewhat different for fourth grade science and mathematics, as can be seen in Exhibit 2.5. For schools with only one mathematics course type, instructional time varied slightly more than mathematics classes for thirteen-year-olds. There was much more variance in the few cases of schools with more than one class type, with medians ranging from 126 to 175 hours annually. Those who offered the least advanced mathematics had the higher (175 hour) median. For these students, who were likely perceived as needing more instructional time, mathematics instruction averaged about 50 minutes per day as compared to 45 minutes per day for those perceived as more able or prepared mathematically.

The range of daily instructional time for nine-year-olds' mathematics was more extensive than the comparable range for thirteen-year-olds. Some children attended schools with an average of less than ten minutes per day devoted

Exhibit 2.4. Percent of Eighth Grade Students in Schools Offering Various Numbers of Science Curricular Types.

	1 Type	2 Types	3 or More Types
USA	80.7 (4.2)	12.4 (2.6)	6.9 (3.0)
Australia	84.2 (2.9)	6.5 (1.9)	9.3 (2.8)
Belgium (Fl)	37.2 (6.2)	22.0 (4.3)	40.8 (5.5)
Canada	92.5 (2.1)	7.2 (2.1)	0.3 (0.3)
Czech Republic	94.1 (3.0)	5.0 (3.0)	0.9 (0.7)
England	77.5 (4.1)	9.1 (3.3)	13.4 (3.9)
France	100.0	0.0	0.0
Germany	90.6 (2.7)	9.4 (2.7)	0.0
Hong Kong	100.0	0.0	0.0
Israel	92.7 (5.2)	3.9 (3.9)	3.5 (3.5)
Japan	100.0	0.0	0.0
Korea	99.3 (0.7)	0.7 (0.7)	0.0
Netherlands	30.1 (5.0)	25.8 (5.5)	44.1 (6.0)
New Zealand	78.1 (3.3)	6.8 (2.0)	15.1 (3.1)
Norway	100.0	0.0	0.0
Russian Federation	94.3 (2.3)	5.7 (2.3)	0.0
Singapore	19.6 (3.9)	80.4 (3.9)	0.0
Spain	100.0	0.0	0.0
Sweden	97.4 (1.5)	1.7 (1.2)	0.9 (0.9)
Switzerland	72.0 (3.4)	14.9 (2.5)	13.1 (2.9)
Thailand	89.4 (4.1)	10.6 (4.1)	0.0

to mathematics, while others attended schools with as much as 120 minutes per day given to mathematics instruction.

From their labels and other data discussed below, the 'class types' for fourth grade mathematics seem clearly to have been based more on ability groups to provide more and less instructional time in an attempt to compensate for differing abilities. This contrasted with thirteen-year-olds' mathematics in which the different class types within a school involved different books and curricular emphases. However, these distinctions in thirteen-year-olds' mathematics – this differential access to educational possibilities through course types and tracking – may also have resulted, at least in part, from a perceived need to provide for different abilities.

Fourth grade science classes commonly had about 100 instructional hours per year and an average of about 35 minutes per day. This was much less than that of eighth grade or that of fourth grade mathematics. The average daily instructional time for nine-year-olds' science ranged from between 10 to 60 minutes per day. It was also much more variable than that of eighth grade, but less so than of fourth grade mathematics.

Exhibit 2.5a *Variation in the Reported Math Instructional Hours for Fourth Grade Students.*

Number of instructional hours/year in math for students and the percent of students following the *least advanced* math course in schools offering more than one math course.

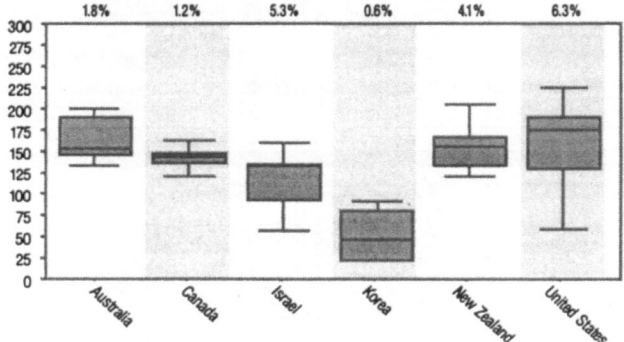

Number of instructional hours/year in math and the percent of students in schools offering only one math course.

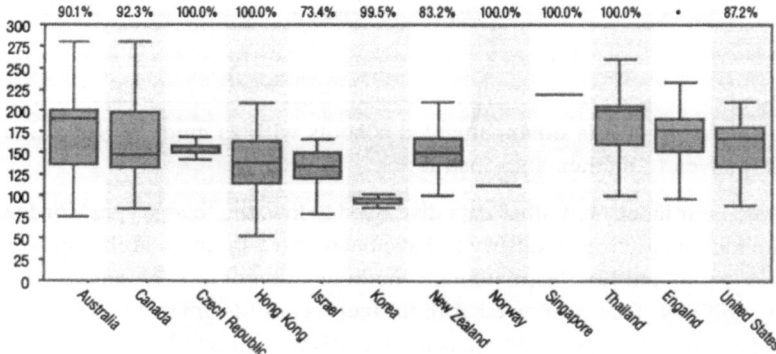

Number of instructional hours/year in math for students and the percent of students following the *most advanced* math course in schools offering more than one math course.

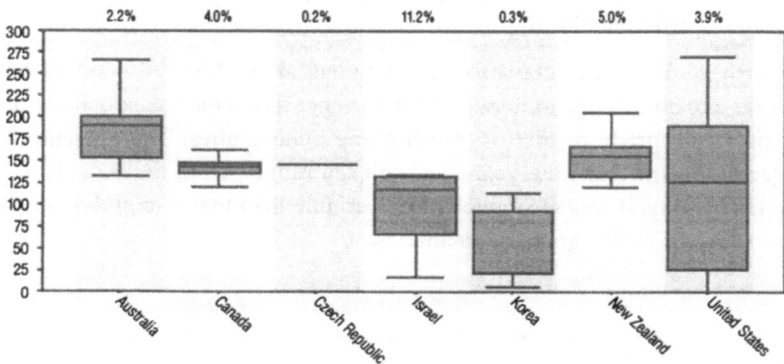

Exhibit 2.5b *Variation in the Reported Science Instructional Hours for Fourth Grade Students.*

Number of instructional hours/year in science for students and the percent of students following the *least advanced* science course in schools offering more than one science course.

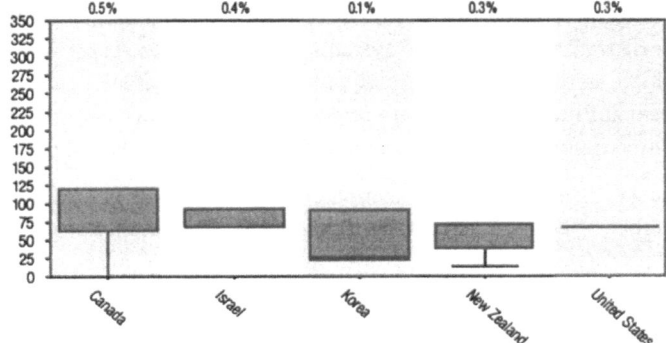

Number of instructional hours/year in science and the percent of students in schools offering only one science course.

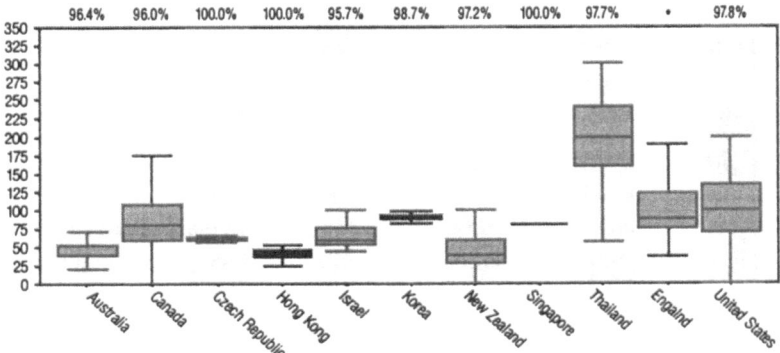

Number of instructional hours/year in science for students and the percent of students following the *most advanced* science course in schools offering more than one science course.

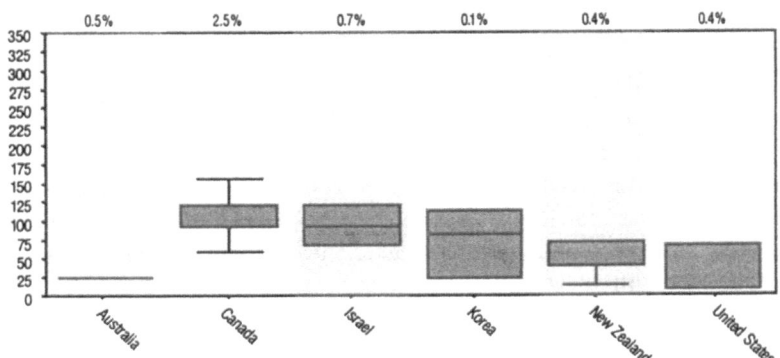

A SUMMARY

Our educational systems' choices have affected mathematics and science instruction even at the broad level of instructional time. The total mathematics and science instructional time in the US was generally higher than that of many other TIMSS countries. It varied according to both grade level, fourth and eighth grade, and academic discipline, science and mathematics. It differed for programs or tracks within those systems and for classes, especially in fourth grade, which were formed apparently, at least in mathematics, to compensate for different abilities. Time allocations within the US showed considerable variability. Three main points summarize most of what these data showed:

- The US is above average compared to other TIMSS countries in the number of hours it devotes to mathematics and science instruction. It seems unnecessary for us to devote more time on average, although those at the lower end of the distribution might want to do so. More time would not necessarily be better if 'more' means 'more of the same thing' we have been doing with the time we have spent until now.

- Within the US, time allocated for mathematics instruction contributed to differences in access to educational possibilities, especially for the nine-year-old population. This differed markedly from the policy context in many other countries in which instructional time was set and held fairly constant among schools and classes.

- Time for access to learning possibilities was not controlled by tracking for US thirteen-year-olds as one might have expected. The instructional time varied somewhat for different class types ('tracks') but was similar for most. For fourth grade students, instructional time was controlled more by ability grouping. The time spent differentiated among these students but the content covered and emphasized, rather than the amount of instructional time, differentiated among eighth grade students as we shall see in the following sections.

WHAT CONTENT DO WE TEACH?

Instructional time is not the only characteristic of what we teach in mathematics and science, nor is it even the most important. Specific areas and topics of mathematics and science (e.g., whole number operations, types of plants, etc.) are involved in the classroom activities that give students access to possibilities for educational experiences. Which specific subject matter contents are chosen is one important characteristic of what we teach. What we ask students to do with these topics is another.

These contents are shaped through educational decision-making dispersed within a US (local or state) educational system. Officials and official bodies (for example, school district administrators, local school boards, etc.) determine instructional goals. They shape what is planned and intended to take place within schools.

An important part of moving from intention to implementation is choosing textbooks. Textbooks are often 'pre-screened' before local decision-makers act. Textbooks are frequently screened by state educational agencies, state textbook committees, and others. Only certain books in either mathematics or science at each grade level are approved for local consideration or listed as allowable for purchase with state funds. Local school administrators, textbook committees, and, in some cases, school boards, further choose among the possible textbooks to be used in the schools for which they are responsible. If that has not resulted in single textbooks for each mathematics or science grade and course, individual school principals and the teachers at that school make the final decision on which textbooks to adopt.

Finally, with official curricula and textbooks in place, teachers must do the work. Official curricula help shape what contents are included and which (if any) are emphasized. Textbooks may help refine this shape by making certain contents and ways of presenting them easier and others more difficult. The long chain of choice begins with the perceptions of officials and policy-makers and ends with teachers' deciding what to teach each day.

Through this often-repeated exercise in dispersed decision-making and interwoven choice, the mathematics and science content of US schools is determined. What we teach and what we emphasize are consequences of those choices. How these choices vary and how they are similar among US educational systems is a product of how these decisions are made. The TIMSS data shed useful light on those similarities and differences.

US CURRICULA: OFFICIAL INTENTIONS, TEXTBOOK IMPLEMENTATIONS

As reported in *A Splintered Vision*, the TIMSS data showed that US mathematics and science curricula were largely unfocused, did not cohere around a small set of powerful mathematics and science ideas that unified those curricula, and were highly repetitive. The data showed that there were indeed contents in common among many of the US' educational systems. They formed a sort of *de facto* definition of what might be considered as 'the basics.' Those 'basics' were not in line with similar 'basics' identified by examining curricula in other TIMSS countries. This was more marked in eighth grade than in fourth grade mathematics (for which this 'core' among US systems was similar to the core of what was common among other TIMSS countries). US fourth grade science curricula, and the resulting composite or 'core' were more focused and aligned with the cross-national composite of all TIMSS countries.

The alignment was extremely good for fourth grade science and still reasonably good in science at the eighth grade level.

One set of numbers serves to illustrate differences in focus between fourth and eighth grades. US textbooks dealt with an average of 55 topics in fourth grade science compared to an average of 68 in eighth grade science.[2] This contrasted with multinational averages of 26 and 30, respectively. The number of topics dealt with was thus considerably larger at eighth grade compared to fourth, a difference not reflected in other TIMSS countries. The differences were achieved by keeping virtually all of the fourth grade topics in eighth grade while adding more topics from physics and chemistry. This is a clear indication of comparatively less focus for eighth grade within the US and, at both grade levels, less focus than was found cross-nationally.

US fourth grade mathematics textbooks worked with an average of 32 topics compared to 38 at the eighth grade level. This contrasted with a cross-national average of 18 and 23, respectively. The differences typically came from retaining the same arithmetic-oriented topics from fourth grade and supplementing them with some algebra and geometry material in eighth grade. Certainly, US textbooks were much less focused at both grade levels than those of other countries and, within the US, somewhat less focused at eighth grade compared to fourth. Arithmetic content continued to dominate US eighth grade textbooks in a way that had no counterpart in most other TIMSS countries. US textbooks for year-long algebra courses in eighth grade mathematics represented no great improvement.

A further examination showed that curricula for the US middle school years were a particular weakness in comparison to other TIMSS countries. Each participating country was asked to indicate, for each science and mathematics topic considered in TIMSS, whether that topic was covered at each grade and whether it was the focus of special instructional attention. There were thus about 80 and 40 topics (science and mathematics, respectively) that could be considered for each of the grades from fifth grade to eighth grade in characterizing topic coverage and focus in the middle school years.

Combining data for science in these four grades showed an average of almost 12 topics introduced with focused attention during these grades for the top achieving countries and an average of more than eight overall among the TIMSS countries. By comparison, the US had no science topics that were introduced with focused attention during this period. For mathematics in these same four grades, the top achieving countries introduced an average of almost eight topics and the overall international average was six.[3] In contrast, the US introduced only one topic with focused attention.

We now turn to an example to illustrate the possible effect of weak middle school curricula on achievement. Exhibit 2.6 on page 28 focuses on only one sub-area of mathematics to provide an example of the decline in performance of eighth grade compared to that of fourth grade. The mean scores for all countries giving the fourth and eighth grade mathematics tests are shown. Note that, in general, fewer countries gave the fourth grade tests (in either science or mathematics). Each test had many geometry items, although those on the fourth grade test were necessarily simpler for the most part than those on the eighth grade test. Each set of geometry items was combined to provide a geometry 'sub-score' for each test. These sub-scores are given here as the average percent of items answered correctly by the students of each country.

The results are divided into three 'tiers' for each test. The top tier represents those countries that scored significantly above the US on the test. The lower tier represents those that scored significantly below the US. The middle tiers are those countries whose scores did not differ significantly from the US. While the US fourth grade students responded to simpler, grade-appropriate geometry items, they averaged a much higher number of items correct than did the US eighth grade students (71 versus 48 percent, respectively). Further, at the fourth grade level, the US scored above the international average in geometry, was in the upper part of the middle tier, and actually was (statistically) significantly outperformed by only two countries (Hong Kong and Australia). By contrast, at eighth grade the US was significantly outperformed by almost 60 percent of the countries and were near the bottom of the country rankings. The US outperformed only Iran, Colombia, Kuwait, and South Africa.

While it is hard to assess the difficulty of the fourth grade items versus the eighth grade items for their respective students, these data suggest strongly that US students at the end of middle school achieved less well than US students at the beginning of middle school. This coincides with a curriculum for those years that must be characterized as not including much focus on geometry. Obviously there are too many other factors involved for such coincidence to be conclusive, but it is one piece of evidence that suggests that US students fall behind over the middle school years. Again this is difficult to assess from these data since the same students were not followed over the four grades, but rather two parallel 'cross-sections' were taken simultaneously to provide an indication of effects over time.

It may be helpful to examine the sort of items to which students were responding in the two geometry tests in order to support this hypothesis further. Exhibit 2.7 shows two geometry items, one each from the nine-year-olds' and thirteen-year-olds' test, along with the data on US and other countries percentages correct (p-values). The item for younger students is a simple item that asks if they can recognize lines of symmetry. They must also be able to inspect several figures visually to identify potential lines of symmetry. About 74 percent of US students in fourth grade and about 64 percent of those in third grade got the item correct, comparing favorably to the average for comparable grades over all TIMSS countries, 64 and 54 percent, respectively.

Exhibit 2.6. *Geometry Achievement for Fourth and Eighth Grade Students Compared to the US (national percent correct).*

Grade 4 Nation	% Correct
Hong Kong	74
(Australia)	74
England	74
Scotland	72
Japan	72
Singapore	72
Korea	72
Canada	72
(Slovenia)	72
(Netherlands)	71
United States ••••••••	71
Czech Republic	71
(Austria)	67
(Latvia(LSS))	67
Ireland	66
New Zealand	66
(Hungary)	66
International	*64*
Iceland	63
(Israel)	62
Norway	58
Greece	53
(Thailand)	53
Cyprus	53
Portugal	52
Iran, Islamic Republic	42
(Kuwait)	36

Grade 8 Nation	% Correct
Japan	80
Singapore	76
Korea	75
Hong Kong	73
Czech Republic	66
France	66
(Bulgaria)	65
Belgium-Flemish	64
Russian Federation	63
Slovak Republic	63
(Thailand)	62
(Slovenia)	60
Hungary	60
Switzerland	60
(Netherlands)	59
(Belgium-French)	58
Canada	58
(Australia)	57
(Israel)	57
(Austria)	57
Latvia (LSS)	57
International	*56*
New Zealand	54
England	54
(Denmark)	54
Lithuania	53
(Romania)	52
(Scotland)	52
Ireland	51
(Germany)	51
Iceland	51
Norway	51
(Greece)	51
Spain	49
Sweden	48
United States ••••••••	48
Cyprus	47
Portugal	44
Iran, Islamic Republic	43
(Kuwait)	38
(Colombia)	29
(South Africa)	24

☐ Significantly Higher
☐ No Significant Difference
▦ Significantly Lower

Source: Mullis, I.V.S., Martin, M.O., Beaton, A.E., Gonzalez, E.J., Kelly, D.L. and Smith, T.A., (1997). *Mathematics Achievement in the Primary School Years: IEA's Third International Mathematics and Science Study*. Chestnut Hill, MA: Center for the Study of Testing, Evaluation, and Educational Policy, Boston College, p.47.

Beaton, A.E., et. al. (1996). *Mathematics Achievement in the Middle School Years: IEA's Third International Mathematics and Science Study*. Chestnut Hill, MA: Center for the Study of Testing, Evaluation, and Educational Policy, Boston College, p.41.

The thirteen-year-olds' geometry test item involves: recalling that 'corresponding parts of congruent triangles are of equal size,' being able to match the two triangles so that corresponding parts are identified with each other (involving transformations at least informally since a triangle must be 'reflected'), recalling that the sum of the size of the angles of a triangle is 180 degrees, and using this information to solve for the size of the third angle. Thus, the item involves two pieces of recall (three, if one counts the idea of a 'corresponding part'), a visual matching skill, some simple arithmetic skills, and enough understanding to put all of these pieces together to obtain a final answer. From this simple 'task analysis' this item seems likely to be only a slightly harder item for thirteen-year-old students than was the nine-year-olds' item for nine-year-old students. Further, for students with four more years of experience and maturation, the slightly more involved performance required to answer this item correctly seems roughly equivalent in difficulty to that required for the nine-year-olds to answer the other item.

Exhibit 2.7. A Comparison of Seventh and Eighth Grade Performance on Two Sample Geometry Items.

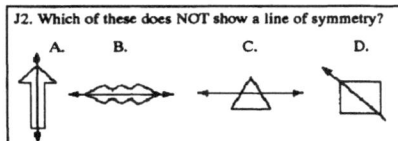

J2. Which of these does NOT show a line of symmetry?	K8. These triangles are congruent. The measure of the sides and angles of the triangles are shown.
A. B. C. D.	What is the value of x? A. 52 B. 55 C. 65 D. 73 E. 75
Item from test for nine-year-olds	Item from test for thirteen-year-olds

	Grade 3		Grade 4	
USA	64.3	(3.0)	73.6	(2.0)
International	54.4		64.0	
Australia	71.6	(3.5)	76.2	(2.3)
Canada	69.7	(3.2)	71.9	(2.5)
Czech Rep.	61.9	(2.4)	73.8	(2.1)
England	68.4	(2.4)	84.0	(2.2)
Hong Kong	82.3	(1.4)	88.7	(1.7)
Hungary	49.7	(2.8)	70.1	(2.8)
Israel			67.8	(3.5)
Japan	42.7	(2.5)	51.5	(2.5)
Korea	77.3	(2.2)	87.2	(1.7)
Netherlands	33.2	(2.8)	40.2	(3.4)
N. Zealand	54.3	(3.3)	57.7	(3.7)
Norway	18.0	(2.8)	29.4	(2.8)
Singapore	55.3	(1.8)	93.4	(0.9)
Thailand	57.9	(3.6)	65.5	(3.0)

	Grade 7		Grade 8	
USA	14.5	(1.8)	17.1	(1.6)
International	27.2		27.2	
Australia	28.7	(2.2)	34.3	(1.8)
Belgium FL	29.5	(2.8)	43.1	(2.8)
Canada	20.4	(2.3)	28.9	(2.5)
Czech Rep.	43.4	(3.7)	50.6	(3.0)
England	24.5	(2.8)	30.5	(3.7)
France	37.8	(3.2)	50.1	(2.8)
Germany	27.7	(2.7)	29.3	(3.0)
Hong Kong	54.7	(3.0)	61.0	(2.7)
Hungary	28.3	(2.4)	39.4	(2.8)
Israel			43.2	(3.4)
Japan	39.9	(2.1)	68.7	(1.7)
Korea	55.4	(2.8)	66.0	(2.1)
Netherlands	14.5	(2.4)	21.2	(3.0)
N. Zealand	18.9	(2.0)	26.1	(2.5)
Norway	24.5	(2.5)	29.8	(2.3)
Russian Fed.	33.0	(3.2)	38.7	(2.9)
Singapore	55.2	(2.8)	69.2	(2.3)
Spain	16.8	(2.0)	13.9	(1.9)
Sweden	18.1	(2.3)	34.2	(2.4)
Switzerland	24.6	(2.1)	32.6	(2.8)
Thailand	22.0	(1.8)	33.0	(2.2)

However, only 17 percent of the thirteen-year-olds got their item correct compared to a cross-national average of 35 percent. Several countries did quite well on this item, including Singapore (69 percent), Japan (69), Korea (66), Hong Kong (66), and France (50). Cross-nationally, this seems to be related to whether or not the content was in the curriculum. The same appears to be true in comparing the performances of US nine- and thirteen-year-olds. At fourth grade, the content needed to answer the item was in the curriculum and a large proportion of students did well. What was 'basic' at eighth grade from the cross-national point of view (and thus what appeared on the test) was not 'basic' in most US eighth grade mathematics curricula. This item was difficult among all TIMSS countries, with a low cross-national percentage correct, most likely because the content was not taught at eighth grade in many countries. However, the US still managed to go from about 10 points above the cross-national average at fourth grade to almost 20 points below at eighth grade.

OFFICIAL INTENTIONS AND TEXTBOOKS: A SUMMARY

We have examined briefly some illustrative data about the official intentions and goals that shape US mathematics and science curricula and about the textbooks which help further to shape their implementation. While we presented only a quick survey, some conclusions seem warranted. These include:

- The US composite mathematics curriculum was unfocused and lacked coherence, especially in the eighth grade. At the fourth grade level, the US composite mathematics curriculum aligned reasonably well with the 'core' or 'composite' curriculum of the other TIMSS countries. At the eighth grade level, there was no similar alignment.

- The US composite science curriculum was more focused, especially at fourth grade. The US composite science curriculum lined up well with the cross-national composite science curriculum at fourth grade and fairly well at eighth grade, although not as well as at the fourth grade level.

- The US middle school curriculum seems to lack intellectual rigor. Certainly reform efforts such as those of the National Council of Teachers of Mathematics *Standards* seek to change this. The indications from the TIMSS data, however, are that neither sufficient implementation nor sufficient improvement had come by the time the data were collected. The current middle school curriculum, especially in mathematics, provides comparatively less access to the intellectual challenges offered students in many other TIMSS countries. Fewer new topics are introduced with focused attention. Unless students are in courses using more challenging texts, what they are taught appears likely to be material covered largely in previous grades. Certainly this seems true in mathematics for eighth grade non-algebra courses. Given the virtual absence of focused introductions of new content in either mathematics or science during fifth through eighth grades, this lack of challenge seems likely. Certainly TIMSS achievement results coincide with what would be expected in this case suggesting a pattern of falling relative standing in both mathematics and science.

US CURRICULA: WHAT OUR TEACHERS TEACH

So far we have examined US mathematics and science curricula as they were reflected in the official intentions of US education systems and in the textbooks chosen to support implementing their intentions. We can also examine what our teachers actually taught according to the TIMSS data. This provides a different picture of curricula in mathematics and the sciences – curricula as they are enacted by teachers in their classroom activities.

How Much US Teachers Teach. Exhibit 2.8 on pages 32 and 33, helps make the point rather dramatically that US teachers typically taught more topics than teachers in other countries. The exhibit has two panels, one for the number of mathematics topics taught by fourth grade teachers and another for the number of science topics taught by those teachers (since most often teachers taught both at this grade level). Within a panel, each row represents the topic coverage of one teacher as they reported it on a TIMSS questionnaire. Each column represents one of the topics about which the teachers were surveyed. The panel for science has more columns than does that for mathematics since teachers were asked about more science topics.

The columns are arranged so that the most commonly taught topics are towards the left of each panel. The rows are arranged so that the teachers who covered the most topics are towards the top of each panel (the row of corresponding rank does not necessarily represent the same teacher in both panels). The line drawn within each panel connects the point in each row representing the total number of topics taught by that row's teacher with similar points for other rows (other teachers).

The result is a visual display of how many topics were taught by a representative sample of fourth grade students' teachers in mathematics and in science. To the extent that the sample was representative, the display for the entire cohort of teachers would show a similar curve but would be 'stretched' vertically since more teachers would be involved. A 'curve' connecting the points in each row that moved quickly from right to left as one's eye moved down the panel would indicate that comparatively few teachers had covered large numbers of topics. A curve that moved more slowly from right to left would indicate that comparatively more teachers had covered large numbers of topics.

Exhibit 2.8 illustrates that most US fourth grade students' teachers covered comparatively many topics for both mathematics and science. The only countries that had somewhat similar patterns to the US were Australia, New Zealand, and, in science, Canada. In the US, both curves stay relatively vertical and move only slowly to the left as one's eye moves from top to bottom in each panel. Only a small portion at the bottom of each panel shows a tapering to markedly fewer topics.

Exhibit 2.8. *Number of Topics Covered by Sampled Fourth Grade Students'*
Mathematics and Science Teachers for a Selected Set of Countries.

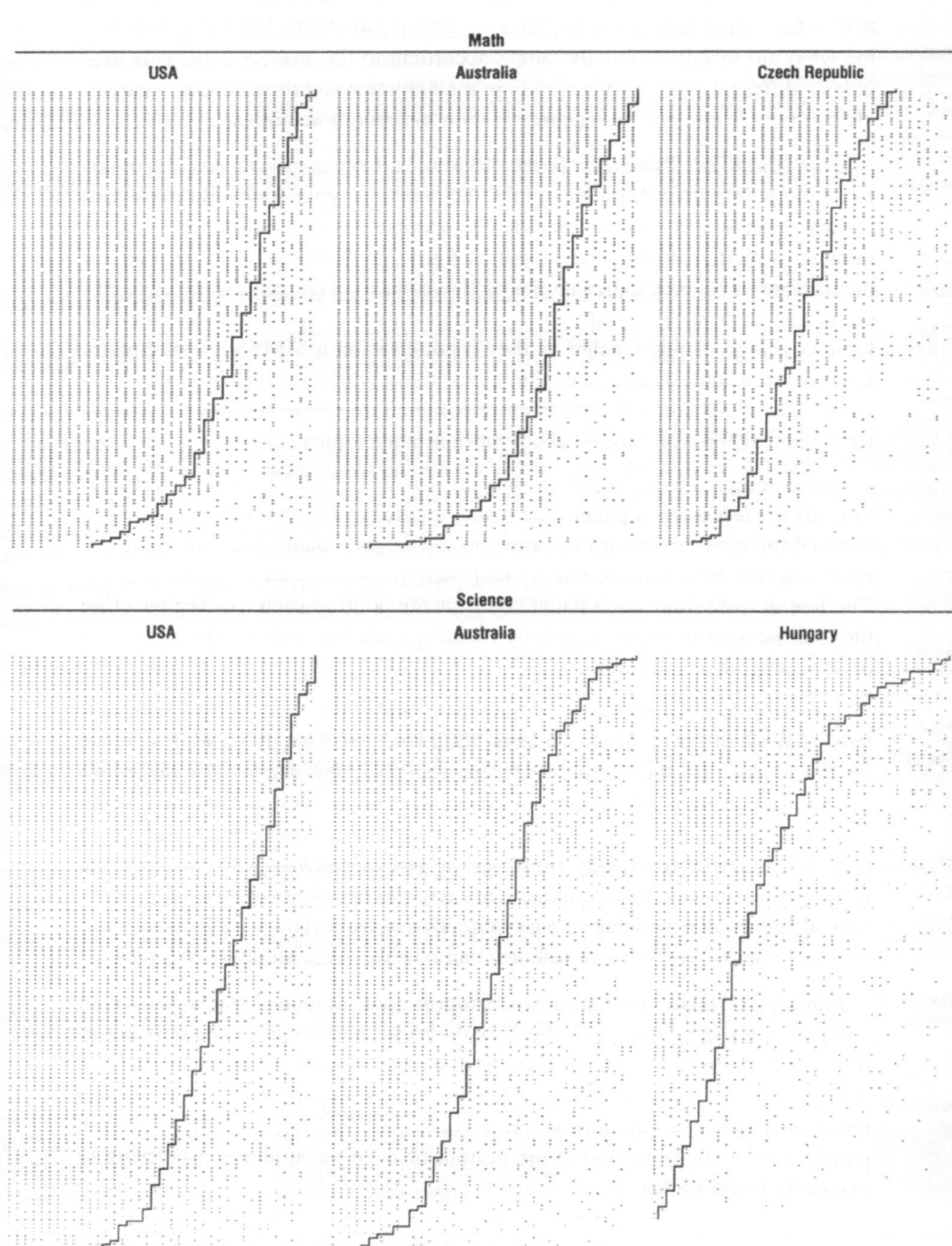

Exhibit 2.8. *(Continued)*

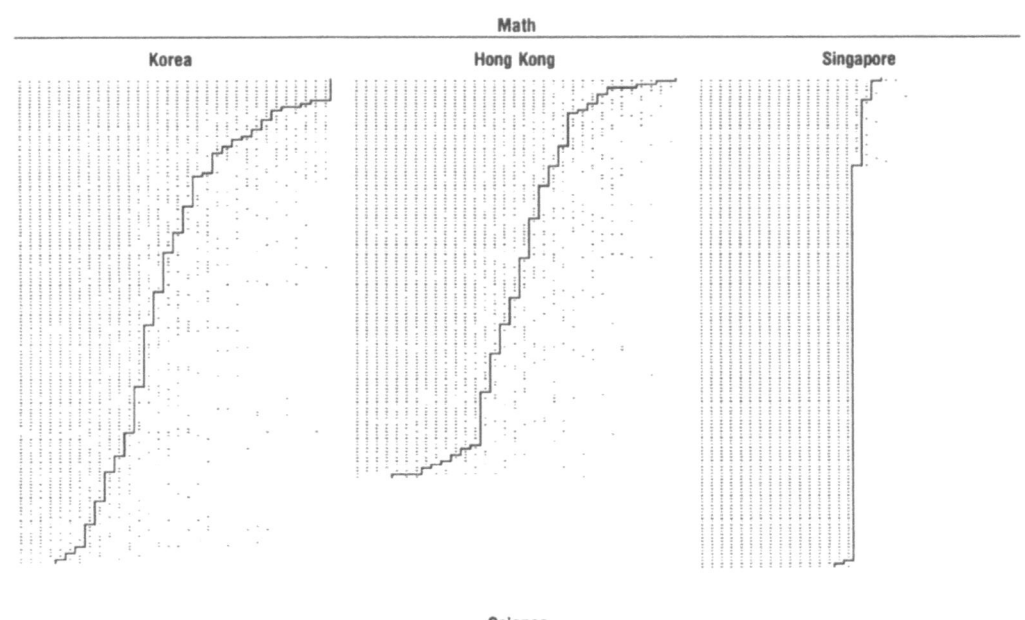

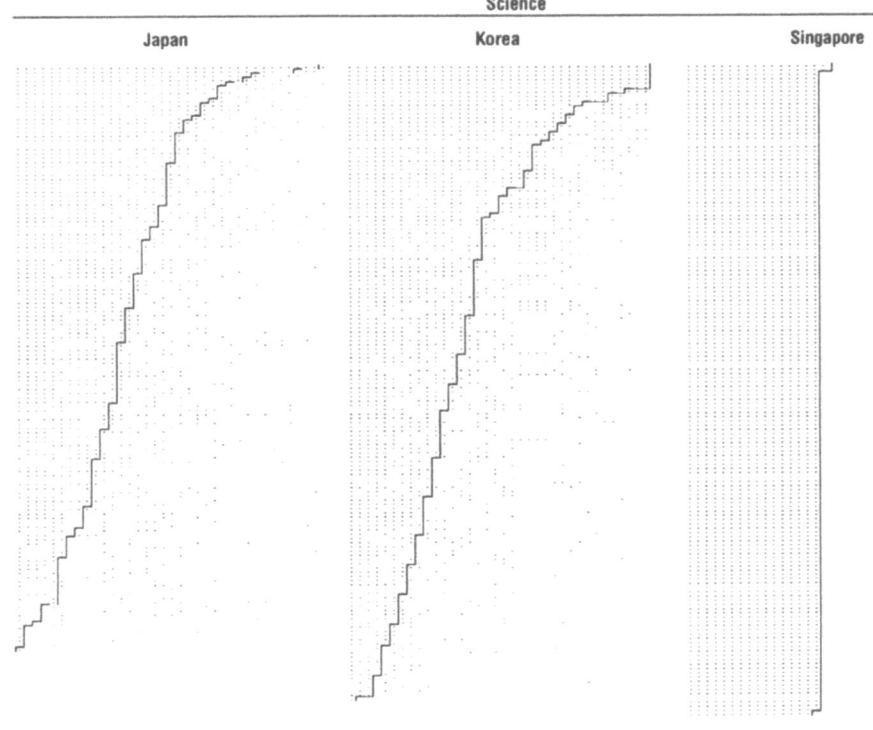

Exhibit 2.9. *Number of Topics Covered by Sampled Eighth Grade Math Teachers in Selected Countries.*

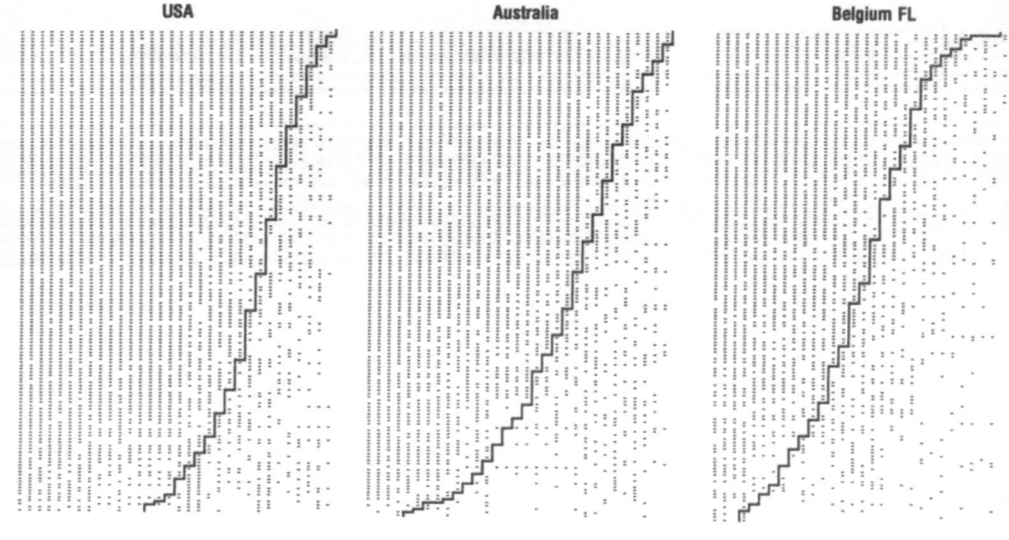

The display for thirteen-year-olds is even more dramatic. Exhibit 2.9 shows comparable displays for mathematics teachers in the higher of the two adjacent grades containing the most thirteen-year-olds in a select set of countries (eighth grade in the US). Clearly the US' curve moves more slowly to the left than does that for most other countries, especially for Korea and others for which the curve moves markedly to the left near the top of the panel. Comparatively more US eighth grade mathematics teachers taught higher numbers of topics (all countries were surveyed about coverage of the same topics). Again, the only countries that had patterns similar to that of the US were Australia and Canada. In contrast, relatively few sampled Japanese teachers covered many topics and most covered far fewer and thus presumably more carefully selected topics (Schmidt, McKnight, & Raizen, p. 70).

In all panels, the 'gaps' in rows to the left of the curve and the 'dashes' to its right indicate that some teachers covered different topics from others. Only the most commonly covered topics (the leftmost columns) were taught by most teachers. Almost every teacher in each country indicated that they did not cover at least some widely covered topics (the gaps) and that they did cover some that were less commonly covered by other teachers in that country (the dashes to the right of the curve).

Clearly, teachers in each country exercised some autonomy in selecting which topics they covered regardless of curriculum policies. In the more

Exhibit 2.9. (Continued).

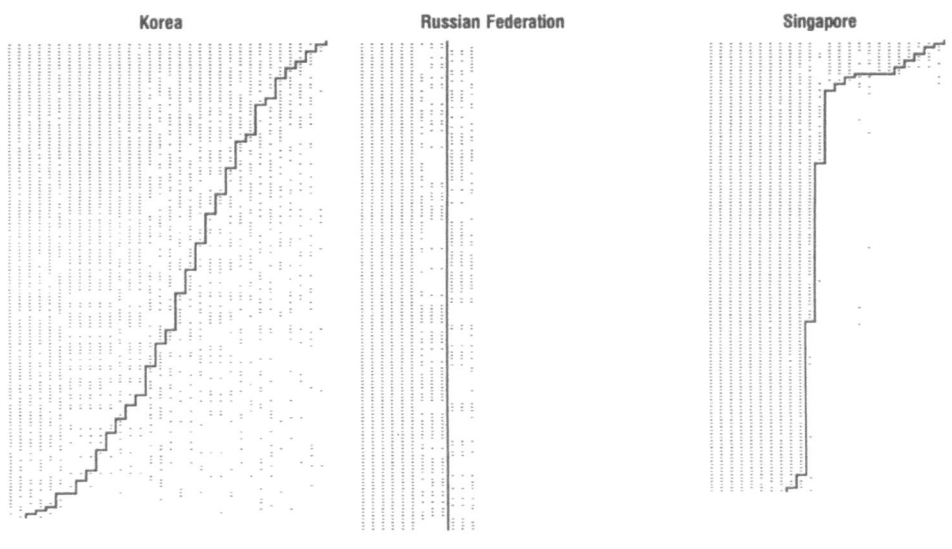

Korea   Russian Federation   Singapore

curricularly-centralized national systems (for example, Korea and Japan), the exhibited gaps and dashes indicate that teachers made some deviations from the official national curriculum. Singapore, another curricularly-centralized system, reflects almost total uniformity in what was taught. Oddly, the splintered US curriculum has almost as much uniformity among teachers as does that of Singapore, but only because virtually all teachers teach almost every topic. US mathematics curricula may intend coverage of a little of everything and US teachers try to deliver it, much as Singapore's teachers do for their more selective curriculum.

The 'movement' of the curves from right to left as one's eye moves down each panel indicates variance in the numbers of topics covered by teachers. Furthermore, sufficiently many topics were on the surveys and sufficiently many teachers sampled to reveal surprising variance in *which* topics were covered and omitted, even allowing for variance in the *number* of topics. Similar variance in the topics covered and omitted is seen in Exhibit 2.8 for US nine-year-olds' teachers, even when there was less variance in the (high) number of topics covered.

Exhibit 2.10 contains the corresponding TIMSS data for science teachers in the upper grade for thirteen-year-olds. Here, the curve for the US still traverses from right to left more slowly than do those for many of the other countries. For some countries, such as the Czech Republic and Korea, the curve moves

Exhibit 2.10. Number of Topics Covered by Sampled Eighth Grade Science Teachers in Selected Countries.

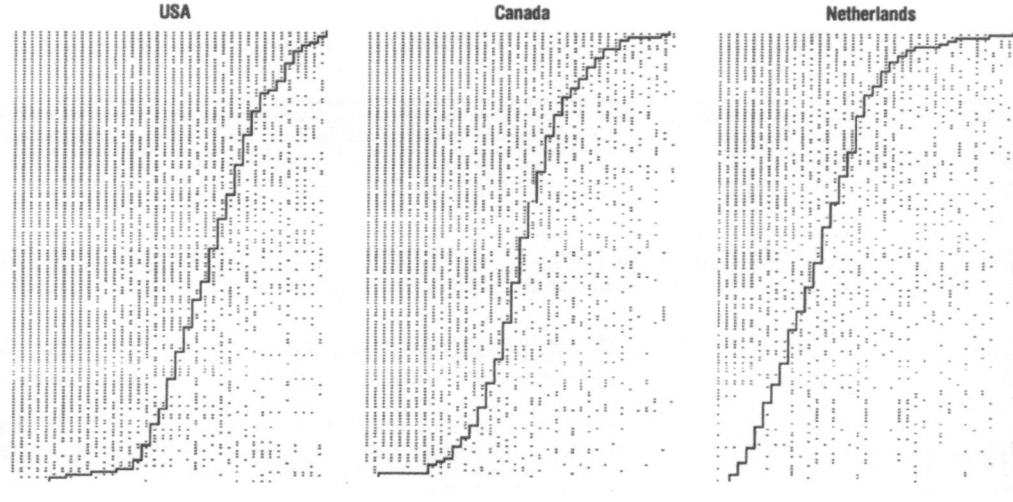

to the left much more quickly for science teachers than in the corresponding curve for mathematics teachers. The US' curve for science teachers moves to the left somewhat more rapidly than the corresponding curve for mathematics teachers. We conclude that US eighth grade science teachers also taught comparatively more topics than did their counterparts in other countries, but not as markedly so as US eighth grade mathematics teachers.

What They Teach. Which topics are most commonly covered by US teachers? Exhibit 2.11 on page 38 examines the specific topics covered by US fourth grade mathematics teachers. If we take as our criterion 'topics covered by 90 percent of the teachers,' the TIMSS data suggest strongly that US fourth grade mathematics curricula as indicated by what teachers taught was unsurprisingly arithmetic-centered, but not exclusively so.[4] The teachers focused on whole numbers, fractions, decimals, and 'estimation and number sense.' They also included measurement topics (units, perimeter, area, etc.), congruence and similarity in simple plane figures, and some aspects of 'data representation and statistics' (most likely averages and using simple chart types). This coincides quite closely with the 'core' or 'common' covered topics among all TIMSS countries – the cross-national 'basics.' In contrast, only two topics reached this level of consensus in US fourth grade science – 'earth features' and 'environmental and resource issues.' There were far fewer 'consensus' or 'core' topics among US fourth grade science teachers than among fourth grade mathematics teachers.

Exhibit 2.12 presents the corresponding data on specific topics covered by US eighth grade mathematics teachers. Using the same 'covered by 90 percent

Exhibit 2.10. (Continued).

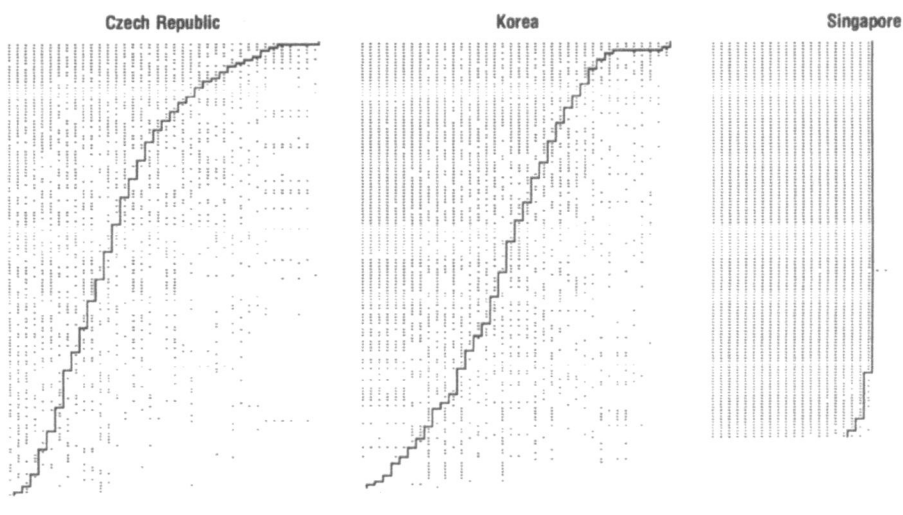

of the teachers' criterion, US eighth grade mathematics 'taught' curricula were surprisingly arithmetic-centered. They contained as widely common topics: common fractions, decimal fractions, percentages, number sets, (simple) number theory (such as greatest common divisors as seen in common denominators), 'estimation and number sense', and ratio and proportion (which, although going beyond simple arithmetic, still has a strong arithmetic element). These topics were supplemented by some measurement ('perimeter, area, and volume') and the 'basics' of one- and two-dimensional geometry (points, lines, angles, etc.).

This was quite different form the algebra- and geometry-centered cross-national consensus curriculum. For instance, functions, aspects of three-dimensional geometry, 'congruence and similarity', and geometric transformations were all part of the cross-national 'core' topics widely covered among TIMSS countries. These topics did not even reach the 75 to 89 percent category for topics covered by US teachers. Instead, they were covered by only 50 to 74 percent of those teachers.

Comparing fourth and eighth grade mathematics topics, many were widely covered by US teachers at both grades. Taking 'taught by at least 75 percent of the teachers at both grades' as our comparison criterion, common topics included whole number arithmetic, fractions, measurement, and basic geometry. If we judged only by the topic names ('functions, relations, and patterns' or geometric 'congruence and similarity'), the topics found in fourth but not eighth grade mathematics often sound more advanced than the eighth grade

Exhibit 2.11. Topics Covered by US Fourth Grade Mathematics Teachers*.

Topics Taught by 90-100% of the Teachers
Whole Numbers
Common and Decimal Fractions
Estimation and Number Sense
Measurement Units and Processes
Perimeter, Area, and Volume
Geometric Congruence and Similarity
Data Representation and Statistics
Topics Taught by 75-89% of the Teachers
Number Theory
1-D and 2-D Geometry Basics
Geometric Transformations and Symmetry
Functions, Relations, and Patterns
Probability
Topics Taught by 50-74% of the Teachers
Estimation and Error of Measurments
Equations and Formulas
Topics Taught by Less Than 50% of the Teachers
Percentages
Number Sets and Concepts
Constructions and 3-D Geometry
Sets and Logic
Ratio and Proportion
Other Advanced Topics

* See note 4 on page 61.

topics. However, as fourth grade content, they most likely represented comparatively simpler aspects of those topics than they would have if the content had been found at the eighth grade level.

The topics covered by at least 75 percent of US eighth grade mathematics teachers, but not fourth grade teachers, included 'number sets and concepts', 'ratio and proportion' and 'equations, inequalities, and formulas.' The 'equations' topic was covered by 50 to 74 percent of US fourth grade mathematics teachers. The other two were covered by less than 50 percent of fourth grade teachers. These contents were most likely in more advanced form in eighth than fourth grade, and certainly were more widely covered. They thus represented some common advances over fourth grade mathematics. However, this was in addition to the arithmetic-centered core previously described, and in spite of some other topics covered (albeit likely in simpler forms) only in fourth grade mathematics.

Exhibit 2.12. *Topics Covered by US Eighth Grade Mathematics Teachers.*

Topics Taught by 90-100% of the Teachers
Common and Decimal Fractions
Percentages
Number Sets and Concepts
Number Theory
Estimation and Number Sense
Perimeter, Area, and Volume
1-D and 2-D Geometry Basics
Ratio and Proportion

Topics Taught by 75-89% of the Teachers
Whole Numbers
Measurement Units and Processes
Equations, Inequalities, and Formulas

Topics Taught by 50-74% of the Teachers
Estimation and Error of Measurments
Geometric Transformations and Symmetry
Geometric Congruence and Similarity
Constructions and 3-D Geometry
Functions, Relations, and Patterns
Statistics and Data
Probability and Uncertainty

Topics Taught by Less Than 50% of the Teachers
Slope, Trigonometry, and Interpolation
Sets and Logic
Other Advanced Topics

The case for science was different. Exhibit 2.13 on page 40 presents the topics covered by US eighth grade science teachers. The only topics covered by at least 75 percent of both fourth and eighth grade science teachers were the 'nature of science' and 'environmental and resource issues.' There were several widely covered topics at fourth but not eighth grade, mainly aspects of biology and geology. At eighth grade, the commonly covered topics not also covered at fourth grade, were mostly from the physical sciences (types and properties of matter, the structure of matter, 'energy types, sources, and conversions', physical changes), and one non-discipline-centered topic, 'science, technology, and society.' Thus, eighth grade science becomes a more physical-science-oriented curriculum, while fourth grade science seems commonly more diverse.

Exhibit 2.14 on page 42 displays the percentage of US and other countries' third and fourth grade teachers teaching the various science topics on which they were surveyed by the TIMSS questionnaires. Along with this display of the percentage of teachers covering the various topics, the exhibit also displays

Exhibit 2.13. Topics Covered by US Fourth and Eighth Grade Science Teachers.

Fourth Grade Science Teachers	Eigthth Grade Science Teachers
Topics Taught by 90-100% of the Teachers	**Topics Taught by 90-100% of the Teachers**
Earth Features	Structure of Matter
Environmental and Resource Issues	Nature of Science
Topics Taught by 75-89% of the Teachers	**Topics Taught by 75-89% of the Teachers**
Weather	Types and Properties of Matter
Earth Processes	Energy Types, Sources, and Conversions
Diversity of Living Things	Physical Changes
Human Health	Science, Technology, and Society
Human Biology	Environmental and Resource Issues
Structure and Function of Living Things	**Topics Taught by 50-74% of the Teachers**
Life Processes and Systems	Earth Features
Life Cycles and Genetics	Earth Processes
Interaction of Living Things	Earth in the Universe
Animal Behavior	Energy Processes
Energy Processes	Chemical Changes
Nature of Science	Special Chemical Changes
Topics Taught by 50-74% of the Teachers	Forces and Motion
Historic Earth Processes	History of Science and Technology
Earth in the Universe	**Topics Taught by Less Than 50% of the Teachers**
Matter	Diversity and Structure of Living Things
Energy Types, Sources, and Conversions	Life Processes and Systems
Physcial and Chemical Changes	Life Cycles, Genetics
Forces and Motion	Interactions of Living Things
History of Science and Technology	Human Biology and Health
Topics Taught by Less Than 50% of the Teachers	Kinetic and Quantum Theory
Science, Technology, and Society	Relativity Theory

the average percentage of time reported as spent on each topic. The focus of the discussion that follows concerns what US teachers taught, although the exhibits also display the same information for teachers in other TIMSS countries. This provides an international comparison for what US teachers taught. The general conclusion from Exhibit 2.14, Exhibit 2.15, Exhibit 2.16, and Exhibit 2.18 from a cross-country perspective is that there is great variability in what teachers taught.

For most topics, some US teachers did not teach them. Only two topics ('environment and resource issues' at third and fourth grade, and 'earth features' at fourth grade) were covered by virtually all (more than 90 percent) of the teachers. On the other hand, some time was spent on each topic when averaged over all teachers. This was true for both third and fourth grades. The time devoted by teachers to any one of the 23 topics surveyed never reached 10 percent except for earth features. The few topics on which teachers spent more time were mainly from the earth and life sciences. Certainly these data support the notion of diverse but 'flitting' fourth grade science curricula as enacted by US teachers.

The pattern differed somewhat for third and fourth grade mathematics. Exhibit 2.15 on page 44 presents data for mathematics comparable to that for science in Exhibit 2.14. The display clearly illustrates that there is more variation among the percentage of US teachers teaching the 20 mathematics topics than there was for the 23 science topics. Again, no topic was completely omitted (either as percentage of teachers teaching it or the percentage of time devoted to it). For mathematics, however, a few key topics (whole numbers, fractions, and 'estimation and number sense') had somewhat more time devoted to them in fourth grade than did other topics. Each of the emphasized topics received on average about two to four weeks' instruction (about 10-20 periods).

At the seventh and eighth grade level, the mathematics topic that received the most attention from US teachers was 'fractions and decimals.' This topic received an average of about two months (40 periods) instructional time. Other 'emphasized' topics received no more than two weeks (10 periods) instructional time. The proportions of time spent by teachers on the various topics were almost identical.

Since there were so few differences in time use, we can combine 'pre-algebra' and 'regular' eighth grade mathematics courses together as 'non-algebra' courses and contrast the time their teachers devoted to various topics with the time devoted by 'algebra' course teachers. This is done in Exhibit 2.17. About 20 percent of eighth grade mathematics students were in what is here classified as 'algebra' courses. The rest were in one version or another of what we are calling 'non-algebra' courses.

There were some notable differences in teacher content coverage among the two course types. Algebra courses spent less time than non-algebra courses on fractions (11 versus 19 percent) and more time on slope, functions, and equations (21 versus 9 percent, respectively, for these latter topics combined). Along with these quantitatively different time allocations, it is likely that there were qualitative differences in how the time on slope, functions, and equations was spent in the two types of courses.

While this appears to speak well for US eighth grade algebra courses, this is in contrast to 27 percent of instructional time that Japanese teachers devoted to slope, functions, and equations, and to 29 percent that they devoted to 'congruence and similarity' and two-dimensional geometry 'basics' (points, lines, rays, angles, parallelism, etc.). The five topics combined accounted for about 56 percent of instructional time in Japan. In the US, even algebra students had these five topics account for only 28 percent of the instructional year. The 80 percent of US eighth graders in 'non-algebra' courses had only 16 percent of the instructional year devoted to these topics (to say nothing of the likely concomitant qualitative differences).

Exhibit 2.14. *Science Topics Covered by Third and Fourth Grade Teachers (percent of teachers covering the topics and percent of time spent).*

Third Grade

	Australia		Canada		Czech Republic		Hong Kong		Hungary		Israel	
	Teacher	Time	Teacher	Time	Teacher	Time	Teacher	Time	Teacher	Time	Teacher	Time
Earth Features												
Weather												
Earth Processes												
Historic Earth Processes												
Earth in the Universe												
Structure & Function of Living Things												
Diversity of Living Things												
Life Processes & Systems												
Life Cycles & Genetics												
Interactions of Living Things												
Animal Behavior												
Human Health												
Human Biology												
Matter												
Energy, Types, Sources, Conversions												
Energy Processes												
Physical & Chemical Changes												
Forces and Motion												
Science, Technology, & Society												
History of Science & Technology												
Environmental & Resource Issues												
Nature of Science												

Fourth Grade

	Australia		Canada		Czech Republic		Hong Kong		Hungary		Israel	
	Teacher	Time	Teacher	Time	Teacher	Time	Teacher	Time	Teacher	Time	Teacher	Time
Earth Features												
Weather												
Earth Processes												
Historic Earth Processes												
Earth in the Universe												
Structure & Function of Living Things												
Diversity of Living Things												
Life Processes & Systems												
Life Cycles & Genetics												
Interactions of Living Things												
Animal Behavior												
Human Health												
Human Biology												
Matter												
Energy, Types, Sources, Conversions												
Energy Processes												
Physical & Chemical Changes												
Forces and Motion												
Science, Technology, & Society												
History of Science & Technology												
Environmental & Resource Issues												
Nature of Science												

Teacher Legend: 0% <10% 10-20% 20-30% 30-40% 40-50% 50-60% 60-70% 70-80% 80-90% >90%

Time Legend: <1% 1-5% 5-10% 10-15% >15%

Exhibit 2.14. (Continued).

Third Grade

Japan		Korea		New Zealand		Norway		Singapore		Thailand		USA		
Teacher	Time	Teacher	Time	Teacher	Time	Teacher	Time	Teacher	Time	Teacher	Time	Teacher	Time	

Categories (rows):
- Earth Features
- Weather
- Earth Processes
- Historic Earth Processes
- Earth in the Universe
- Structure & Function of Living Things
- Diversity of Living Things
- Life Processes & Systems
- Life Cycles & Genetics
- Interactions of Living Things
- Animal Behavior
- Human Health
- Human Biology
- Matter
- Energy, Types, Sources, Conversions
- Energy Processes
- Physical & Chemical Changes
- Forces and Motion
- Science, Technology, & Society
- History of Science & Technology
- Environmental & Resource Issues
- Nature of Science

Fourth Grade

Japan		Korea		New Zealand		Norway		Singapore		Thailand		USA		
Teacher	Time	Teacher	Time	Teacher	Time	Teacher	Time	Teacher	Time	Teacher	Time	Teacher	Time	

Categories (rows):
- Earth Features
- Weather
- Earth Processes
- Historic Earth Processes
- Earth in the Universe
- Structure & Function of Living Things
- Diversity of Living Things
- Life Processes & Systems
- Life Cycles & Genetics
- Interactions of Living Things
- Animal Behavior
- Human Health
- Human Biology
- Matter
- Energy, Types, Sources, Conversions
- Energy Processes
- Physical & Chemical Changes
- Forces and Motion
- Science, Technology, & Society
- History of Science & Technology
- Environmental & Resource Issues
- Nature of Science

Teacher Legend: 0% <10% 10-20% 20-30% 30-40% 40-50% 50-60% 60-70% 70-80% 80-90% >90%

Time Legend: <1% 1-5% 5-10% 10-15% >15%

Exhibit 2.15. *Mathematics Topics Covered by Third And Fourth Grade Teachers (percent of teachers covering the topics and percent of time spent).*

Third Grade

	Australia		Canada		Czech Republic		Hong Kong		Hungary		Israel	
	Teacher	Time	Teacher	Time	Teacher	Time	Teacher	Time	Teacher	Time	Teacher	Time
Meaning of Whole Numbers												
Common & Decimal Fractions												
Percentages												
Number Sets & Concepts												
Number Theory												
Estimation & Number Sense												
Measurement Units												
Perimeter, Area & Volume												
Measurement Estimation & Error												
1-D & 2-D Geometry Basics												
Symmetry & Transformations												
Congruence & Similarity												
Constructions & 3-D Geometry												
Ratio & Proportion												
Functions, Relations, Patterns												
Equations & Formulas												
Data & Statistics												
Probability & Uncertainty												
Sets & Logic												
Other Advanced Content												

Fourth Grade

	Australia		Canada		Czech Republic		Hong Kong		Hungary		Israel	
	Teacher	Time	Teacher	Time	Teacher	Time	Teacher	Time	Teacher	Time	Teacher	Time
Meaning of Whole Numbers												
Common & Decimal Fractions												
Percentages												
Number Sets & Concepts												
Number Theory												
Estimation & Number Sense												
Measurement Units												
Perimeter, Area & Volume												
Measurement Estimation & Error												
1-D & 2-D Geometry Basics												
Symmetry & Transformations												
Congruence & Similarity												
Constructions & 3-D Geometry												
Ratio & Proportion												
Functions, Relations, Patterns												
Equations & Formulas												
Data & Statistics												
Probability & Uncertainty												
Sets & Logic												
Other Advanced Content												

Teacher Legend: 0% <10% 10-20% 20-30% 30-40% 40-50% 50-60% 60-70% 70-80% 80-90% >90%

Time Legend: <1% 1-5% 5-10% 10-15% >15%

Exhibit 2.15. (Continued).

Third Grade

Japan	Korea	New Zealand	Norway	Singapore	Thailand	USA	
Teacher Time	Teacher Time	Teacher Time	Teacher Time	Teacher Time	Teacher Time	Teacher Time	

- Meaning of Whole Numbers
- Common & Decimal Fractions
- Percentages
- Number Sets & Concepts
- Number Theory
- Estimation & Number Sense
- Measurement Units
- Perimeter, Area & Volume
- Measurement Estimation & Error
- 1-D & 2-D Geometry Basics
- Symmetry & Transformations
- Congruence & Similarity
- Constructions & 3-D Geometry
- Ratio & Proportion
- Functions, Relations, Patterns
- Equations & Formulas
- Data & Statistics
- Probability & Uncertainty
- Sets & Logic
- Other Advanced Content

Fourth Grade

Japan	Korea	New Zealand	Norway	Singapore	Thailand	USA	
Teacher Time	Teacher Time	Teacher Time	Teacher Time	Teacher Time	Teacher Time	Teacher Time	

- Meaning of Whole Numbers
- Common & Decimal Fractions
- Percentages
- Number Sets & Concepts
- Number Theory
- Estimation & Number Sense
- Measurement Units
- Perimeter, Area & Volume
- Measurement Estimation & Error
- 1-D & 2-D Geometry Basics
- Symmetry & Transformations
- Congruence & Similarity
- Constructions & 3-D Geometry
- Ratio & Proportion
- Functions, Relations, Patterns
- Equations & Formulas
- Data & Statistics
- Probability & Uncertainty
- Sets & Logic
- Other Advanced Content

Teacher Legend: 0% <10% 10-20% 20-30% 30-40% 40-50% 50-60% 60-70% 70-80% 80-90% >90%

Time Legend: <1% 1-5% 5-10% 10-15% >15%

Exhibit 2.16. *Mathematics topics covered by seventh and eighth grade teachers (percent of teachers covering the topics and percent of time spent).*

Seventh Grade

	Australia		Belgium (Fl)		Canada		Czech Republic		France		Germany		Hong Kong	
	Teacher	Time	Teacher	Time	Teacher	Time	Teacher	Time	Teacher	Time	Teacher	Time	Teacher	Time
Meaning of Whole Numbers														
Common & Decimal Fractions														
Percentages														
Number Sets & Concepts														
Number Theory														
Estimation & Number Sense														
Measurement Units														
Perimeter, Area & Volume														
Measurement Estimation & Error														
1-D & 2-D Geometry Basics														
Symmetry & Transformations														
Congruence & Similarity														
Constructions & 3-D Geometry														
Ratio & Proportion														
Slope & Trigonometry														
Functions, Relations, Patterns														
Equations & Formulas														
Data & Statistics														
Probability & Uncertainty														
Sets & Logic														
Other Advanced Content														

Eighth Grade

	Australia		Belgium (Fl)		Canada		Czech Republic		France		Germany		Hong Kong	
	Teacher	Time	Teacher	Time	Teacher	Time	Teacher	Time	Teacher	Time	Teacher	Time	Teacher	Time
Meaning of Whole Numbers														
Common & Decimal Fractions														
Percentages														
Number Sets & Concepts														
Number Theory														
Estimation & Number Sense														
Measurement Units														
Perimeter, Area & Volume														
Measurement Estimation & Error														
1-D & 2-D Geometry Basics														
Symmetry & Transformations														
Congruence & Similarity														
Constructions & 3-D Geometry														
Ratio & Proportion														
Slope & Trigonometry														
Functions, Relations, Patterns														
Equations & Formulas														
Data & Statistics														
Probability & Uncertainty														
Sets & Logic														
Other Advanced Content														

Teacher Legend: 0% <10% 10-20% 20-30% 30-40% 40-50% 50-60% 60-70% 70-80% 80-90% >90%

Time Legend: <1% 1-5% 5-10% 10-15% >15%

Exhibit 2.16. (Continued).

Seventh Grade

Hungary	Japan	Korea	Netherlands	New Zealand	Norway	
Teacher Time	Teacher Time	Teacher Time	Teacher Time	Teacher Time	Teacher Time	

Meaning of Whole Numbers
Common & Decimal Fractions
Percentages
Number Sets & Concepts
Number Theory
Estimation & Number Sense
Measurement Units
Perimeter, Area & Volume
Measurement Estimation & Error
1-D & 2-D Geometry Basics
Symmetry & Transformations
Congruence & Similarity
Constructions & 3-D Geometry
Ratio & Proportion
Slope & Trigonometry
Functions, Relations, Patterns
Equations & Formulas
Data & Statistics
Probability & Uncertainty
Sets & Logic
Other Advanced Content

Eighth Grade

Hungary	Israel	Japan	Korea	Netherlands	New Zealand	Norway	
Teacher Time	Teacher Time	Teacher Time	Teacher Time	Teacher Time	Teacher Time	Teacher Time	

Meaning of Whole Numbers
Common & Decimal Fractions
Percentages
Number Sets & Concepts
Number Theory
Estimation & Number Sense
Measurement Units
Perimeter, Area & Volume
Measurement Estimation & Error
1-D & 2-D Geometry Basics
Symmetry & Transformations
Congruence & Similarity
Constructions & 3-D Geometry
Ratio & Proportion
Slope & Trigonometry
Functions, Relations, Patterns
Equations & Formulas
Data & Statistics
Probability & Uncertainty
Sets & Logic
Other Advanced Content

Teacher Legend	0%	<10%	10-20%	20-30%	30-40%	40-50%	50-60%	60-70%	70-80%	80-90%	>90%

Time Legend	<1%	1-5%	5-10%	10-15%	>15%

Exhibit 2.16. *(Continued).*

Seventh Grade

	Singapore		Spain		Sweden		Switzerland		Thailand		USA	
	Teacher	Time	Teacher	Time	Teacher	Time	Teacher	Time	Teacher	Time	Teacher	Time
Meaning of Whole Numbers												
Common & Decimal Fractions												
Percentages												
Number Sets & Concepts												
Number Theory												
Estimation & Number Sense												
Measurement Units												
Perimeter, Area & Volume												
Measurement Estimation & Error												
1-D & 2-D Geometry Basics												
Symmetry & Transformations												
Congruence & Similarity												
Constructions & 3-D Geometry												
Ratio & Proportion												
Slope & Trigonometry												
Functions, Relations, Patterns												
Equations & Formulas												
Data & Statistics												
Probability & Uncertainty												
Sets & Logic												
Other Advanced Content												

Eighth Grade

	Russian Federation		Singapore		Spain		Sweden		Switzerland		Thailand		USA	
	Teacher	Time	Teacher	Time	Teacher	Time	Teacher	Time	Teacher	Time	Teacher	Time	Teacher	Time
Meaning of Whole Numbers														
Common & Decimal Fractions														
Percentages														
Number Sets & Concepts														
Number Theory														
Estimation & Number Sense														
Measurement Units														
Perimeter, Area & Volume														
Measurement Estimation & Error														
1-D & 2-D Geometry Basics														
Symmetry & Transformations														
Congruence & Similarity														
Constructions & 3-D Geometry														
Ratio & Proportion														
Slope & Trigonometry														
Functions, Relations, Patterns														
Equations & Formulas														
Data & Statistics														
Probability & Uncertainty														
Sets & Logic														
Other Advanced Content														

Teacher Legend: 0% <10% 10-20% 20-30% 30-40% 40-50% 50-60% 60-70% 70-80% 80-90% >90%

Time Legend: <1% 1-5% 5-10% 10-15% >15%

Exhibit 2.17. *US Eighth Grade Mathematics Teacher Coverage of 21 Topics in Algebra and Non-Algebra Courses.*

	Algebra			Non-Algebra		
	Percent of Teachers Teaching	Average Number of Periods	Average Proportion of Time	Percent of Teachers Teaching	Average Number of Periods	Average Proportion of Time
Whole Numbers						
Common & Decimal Fractions						
Percentages						
Number Sets & Concepts						
Number Theory						
Estimation & Number Sense						
Measurement Units						
Perimeter, Area, Volume						
Estimation & Error of Measurements						
1D & 2D Geometry Basics						
Geometric Transformations						
Geometric Congruence & Similarity						
3D Geometry						
Ratio & Proportion						
Slope, Trig. & Interpolation						
Functions, Relations, Patterns						
Equations & Formulas						
Statistics & Data						
Probability & Uncertainty						
Sets & Logic						
Other Advanced Content						

Teacher Legend: 0% 1-10% 10-20% 20-30% 30-40% 40-50% 50-60% 60-70% 70-80% 80-90% 90-100%

Period Legend: <5 5-9 10-15 >15   Time Legend: 0% 1-4.9% 5-9.9% 10-15% >15%

There are some noteworthy patterns in teacher coverage of broad areas of mathematics (such as arithmetic and algebra), especially as we compare third and fourth grades versus seventh and eighth grades for the US and other countries. For arithmetic topics the US does not vary much from seventh and eighth grades as compared to third and fourth grades. Japan and several of the other countries vary significantly. For example, the pattern in Japan and Korea seems to be to cover basic arithmetic content early with third and fourth graders and then to move on by seventh and eighth grade to more algebra- and geometry-centered content. In contrast, in the US, the earlier arithmetic content lingers into seventh and eighth grades producing the pattern of relatively similar coverage of arithmetic topics in the four grades.

US science teachers differed on what was taught and for how long in seventh grade as compared to eighth grade. Exhibit 2.18 displays these data comparing science coverage in the two grades. The topics to which seventh grade science teachers devoted more attention were all from the life sciences ('diversity and structure of living things', 'life processes and systems', and 'human biology and health'). These three topics together received on average more than a third of the year's science instruction. Each of these topics was taught by over 80 percent of the seventh grade science teachers.

Exhibit 2.18. *Science Topics Covered by Seventh and Eighth Grade Teachers (percent of teachers covering the topics and percent of time spent).*

Seventh Grade

	Australia		Belgium (Fl)		Canada		Czech Republic		France		Germany		Hong Kong	
	Teacher	Time	Teacher	Time	Teacher	Time	Teacher	Time	Teacher	Time	Teacher	Time	Teacher	Time
Earth Features														
Earth Processes														
Earth in the Universe														
Diversity & Structure of Living Things														
Life Processes & Systems														
Life Cycles & Genetics														
Interactions of Living Things														
Human Biology & Health														
Types & Properties of Matter														
Structure of Matter														
Energy, Types, Sources, Conversions														
Energy Processes														
Physical Changes														
Kinetic & Quantum Theory														
Chemical Changes														
Special Chemical Changes														
Forces and Motion														
Relativity Theory														
Science, Technology, & Society														
History of Science & Technology														
Environmental & Resource Issues														
Nature of Science														

Eighth Grade

	Australia		Belgium (Fl)		Canada		Czech Republic		France		Germany		Hong Kong	
	Teacher	Time	Teacher	Time	Teacher	Time	Teacher	Time	Teacher	Time	Teacher	Time	Teacher	Time
Earth Features														
Earth Processes														
Earth in the Universe														
Diversity & Structure of Living Things														
Life Processes & Systems														
Life Cycles & Genetics														
Interactions of Living Things														
Human Biology & Health														
Types & Properties of Matter														
Structure of Matter														
Energy, Types, Sources, Conversions														
Energy Processes														
Physical Changes														
Kinetic & Quantum Theory														
Chemical Changes														
Special Chemical Changes														
Forces and Motion														
Relativity Theory														
Science, Technology, & Society														
History of Science & Technology														
Environmental & Resource Issues														
Nature of Science														

Teacher Legend: 0%   <10%   10-20%   20-30%   30-40%   40-50%   50-60%   60-70%   70-80%   80-90%   >90%

Time Legend: <1%   1-5%   5-10%   10-15%   >15%

Exhibit 2.18. *(Continued).*

Seventh Grade

Hungary	Japan	Korea	Netherlands	New Zealand	Norway	
Teacher Time	Teacher Time	Teacher Time	Teacher Time	Teacher Time	Teacher Time	

- Earth Features
- Earth Processes
- Earth in the Universe
- Diversity & Structure of Living Things
- Life Processes & Systems
- Life Cycles & Genetics
- Interactions of Living Things
- Human Biology & Health
- Types & Properties of Matter
- Structure of Matter
- Energy, Types, Sources, Conversions
- Energy Processes
- Physical Changes
- Kinetic & Quantum Theory
- Chemical Changes
- Special Chemical Changes
- Forces and Motion
- Relativity Theory
- Science, Technology, & Society
- History of Science & Technology
- Environmental & Resource Issues
- Nature of Science

Eighth Grade

Hungary	Israel	Japan	Korea	Netherlands	New Zealand	Norway	
Teacher Time	Teacher Time	Teacher Time	Teacher Time	Teacher Time	Teacher Time	Teacher Time	

- Earth Features
- Earth Processes
- Earth in the Universe
- Diversity & Structure of Living Things
- Life Processes & Systems
- Life Cycles & Genetics
- Interactions of Living Things
- Human Biology & Health
- Types & Properties of Matter
- Structure of Matter
- Energy, Types, Sources, Conversions
- Energy Processes
- Physical Changes
- Kinetic & Quantum Theory
- Chemical Changes
- Special Chemical Changes
- Forces and Motion
- Relativity Theory
- Science, Technology, & Society
- History of Science & Technology
- Environmental & Resource Issues
- Nature of Science

Teacher Legend: 0% <10% 10-20% 20-30% 30-40% 40-50% 50-60% 60-70% 70-80% 80-90% >90%

Time Legend: <1% 1-5% 5-10% 10-15% >15%

Exhibit 2.18. (Continued).

Seventh Grade

	Singapore	Spain	Sweden	Switzerland	Thailand	USA
	Teacher Time	Teacher Time	Teacher Time	Teacher Time	Teacher Time	Teacher Time
Earth Features						
Earth Processes						
Earth in the Universe						
Diversity & Structure of Living Things						
Life Processes & Systems						
Life Cycles & Genetics						
Interactions of Living Things						
Human Biology & Health						
Types & Properties of Matter						
Structure of Matter						
Energy, Types, Sources, Conversions						
Energy Processes						
Physical Changes						
Kinetic & Quantum Theory						
Chemical Changes						
Special Chemical Changes						
Forces and Motion						
Relativity Theory						
Science, Technology, & Society						
History of Science & Technology						
Environmental & Resource Issues						
Nature of Science						

Eighth Grade

	Russian Federation	Singapore	Spain	Sweden	Switzerland	Thailand	USA
	Teacher Time	Teacher Time	Teacher Time	Teacher Time	Teacher Time	Teacher Time	Teacher Time
Earth Features							
Earth Processes							
Earth in the Universe							
Diversity & Structure of Living Things							
Life Processes & Systems							
Life Cycles & Genetics							
Interactions of Living Things							
Human Biology & Health							
Types & Properties of Matter							
Structure of Matter							
Energy, Types, Sources, Conversions							
Energy Processes							
Physical Changes							
Kinetic & Quantum Theory							
Chemical Changes							
Special Chemical Changes							
Forces and Motion							
Relativity Theory							
Science, Technology, & Society							
History of Science & Technology							
Environmental & Resource Issues							
Nature of Science							

Teacher Legend: 0% <10% 10-20% 20-30% 30-40% 40-50% 50-60% 60-70% 70-80% 80-90% >90%

Time Legend: <1% 1-5% 5-10% 10-15% >15%

This most likely reflects the fact that most seventh grade classes in the US were life science classes. Exhibit 2.19 shows the percentage of classes at the seventh and eighth grades that were classified as life, earth, physical, and general science. Notice that eighth grade science courses were somewhat evenly divided between general, physical, and earth sciences, with a much smaller proportion for life science. In contrast, for seventh grade almost 60 percent of the courses were life science, followed by another 30 percent general science and far smaller proportions for physical and earth sciences.

Exhibit 2.19. *Percents of US Seventh and Eighth Grade Students in Each Type of Science Course.*

	Seventh Grade	Eighth Grade
General	29.6 (5.2)	28.4 (3.0)
Physical	4.0 (2.0)	29.2 (4.8)
Life	56.7 (5.3)	7.7 (2.7)
Earth	9.7 (3.7)	34.7 (4.2)

Eighth grade science teachers distributed instructional time more evenly among topics. No set of three topics (to provide a comparison to the three seventh grade topics above) accounted for more than 30 percent of instructional time. The four most heavily emphasized topics were 'earth features', the structure, types and properties of matter, and the 'earth in the universe.' This likely reflects the fact that earth and physical science courses were more common for US eighth graders.

The emphases of 'taught' curricula for both seventh and eighth grade science varied among the four main course types (see Exhibit 2.20 for the data for the eighth grade). Each emphasized either a particular area of science or else 'general science.' For seventh and eighth grade science courses focusing on earth science, three topics ('earth features', 'earth processes', and 'earth in the universe') accounted for almost half of the instructional time. The rest of the instructional time typically was spread over biology, physics, and chemistry topics.

Courses oriented to the life sciences demonstrated little difference in focus for the two grades. For seventh grade, the three life science topics mentioned earlier accounted for about 45 percent of instructional time according to teachers' reports. At eighth grade, these same three topics accounted for about 44 percent of instructional time. The rest of the content in these courses was spread over earth and physical science topics.

Courses oriented to the physical sciences emphasized 'types and properties of matter', the 'structure of matter', 'energy types and sources', 'energy

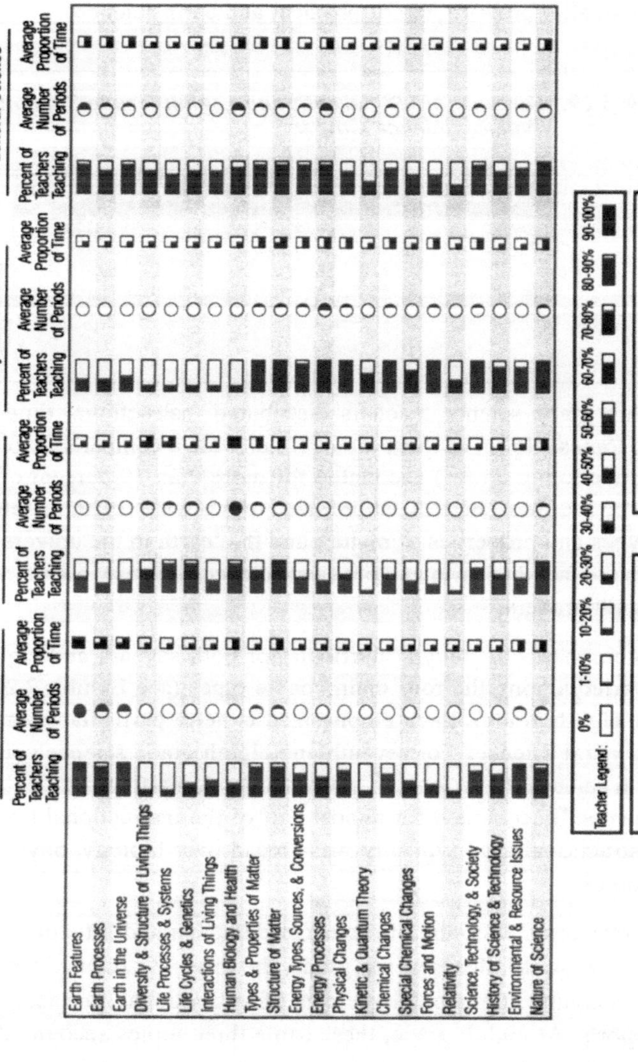

Exhibit 2.20. US Eighth Grade Teacher Coverage of 22 Science Topics in Each Type of Science Course.

processes', 'physical changes', and 'forces and motion.' At seventh grade, these topics accounted for 59 percent of the instructional year in physical-science-oriented courses. At eighth grade they accounted for 54 percent of instructional time and 'chemical changes' accounted for an additional 8 percent. The remainder of the time was spread over life and earth science topics.

'General science' courses, in contrast to courses with a disciplinary focus, had little focus on any small set of topics as indicated by teachers reports. Seventh grade 'general science' teachers focused slightly more on topics in the life sciences. Eighth grade 'general science' teachers devoted less than10 percent of the instructional year to any given topic. That is, there was essentially no focus in the 'taught curricula' for eighth grade general science courses.

In comparison, eighth grade Japanese science teachers reported focusing on 'life processes and systems', 'human biology and health', the 'types and properties of matter', 'structure of matter', 'energy processes', and 'chemical changes.' While we must grant that this was six of 22 possible topics, this (roughly) one-fourth of the topics accounted for 69 percent of the instructional year. When a fourth of the possible topics accounts for more than three-fifths of a year's instruction, we must consider this as an example of comparative focus in a 'taught' curriculum, even when the topics come from both physical and life sciences rather than constituting a course oriented to a single discipline. This is true in comparison to US discipline-oriented courses for which almost no small set of topics approached Japan's reported 69 percent. It is especially true in comparison to the US general science courses for which teachers reported devoting no more than 10 percent of instructional time to any topic.

WHAT OUR TEACHERS TEACH: A SUMMARY

- These results suggest that, to a large extent, we 'got what we intended to get' from US teachers. Curricular goals and intentions for science and mathematics in US educational systems lacked focus as they were reflected in official documents and textbooks. US science and mathematics teachers for nine- and thirteen-year-old students responded to these diverse intentions by covering a large, diverse, and unfocused collection of topics. It was the exception, not the rule, for a few topics to dominate the time that teachers devoted to science or mathematics instruction.

- The composite or 'core' topics – the 'empirical basics' – widely taught by US fourth grade teachers matched well with the cross-national core of what was widely taught in TIMSS countries. Unlike many of the other countries, US teaching (as did US curricula and textbooks) included a larger 'shell' of other topics around this shared core than was typical for other countries. This was especially true in mathematics.

- The US's 'taught' curricula for eighth grade mathematics was unfocused, broad, and redundantly covered many topics found in previous grades. The US composite mathematics curriculum, as reflected by teachers' time allocations, was more arithmetic-centered compared to a more algebra- and geometry- centered cross-national composite.

- There was considerable redundancy between what US fourth grade and US eighth grade mathematics teachers reported that they taught. This was further evidence that US mathematics curricula were not only 'a mile wide and an inch deep' but also highly given to redundancy and review.

- There were no real differences in content coverage reported by eighth grade teachers between 'pre-algebra' and 'regular' or 'general' mathematics courses. This was true even when US schools reported different course types and titles, different textbooks, and officially different syllabi.

- The 'algebra' option in the US eighth grade did differ in content coverage from the other course types in terms of what teachers reported they taught (as well as in course type and title, syllabi, and textbooks). However, the US 'algebra' option was far different from the corresponding Japanese 'algebra' option. The US algebra option still included a breadth of topics as well as a central focus on equations. It had no counterpart to the Japanese second focus on geometry.

- US seventh grade science was mostly biology (when judged by teachers' reports of what they taught). US eighth grade science seemed to be a mixture of earth and physical sciences, even when a course type supposedly emphasized one of those disciplines over the other.

- US science instruction, although officially focused on one particular area (for example, life sciences) was not focused as delivered by teachers. About one-half of the instructional year focused on topics in the targeted area or discipline. The rest – the content delivered for the other half of the instructional year – came from the other three disciplines of science (physical, chemistry, and earth science).

- Both US mathematics and science courses, especially for the eighth grade, were officially characterized by different course types, although this was only partially reflected in what teachers covered, emphasized, or omitted. Mathematics displayed course types based more on how advanced students were in their exposure to (and, presumably, achievement in) mathematics at the time of selecting their current mathematics course. They were, in short, a reflection of assigning students to different tracks that largely determined their access to educational possibilities in mathematics.

- In contrast, science courses seemed to have been based on intentions to focus on different disciplines – although those attempts at focus were typically less marked in the content US science teachers reported covering. Even with this different basis, however, the science course types tended to differentiate access to educational possibilities in science within the eighth grade.

WHO IS TAUGHT WHAT?

Who gets the chance to learn what? Who gains access to different educational possibilities in mathematics and science? The American ideological answer is 'whoever wants to' gets the chance to learn whatever our schools can offer them – that anyone can have full access to all our educational possibilities. The American empirical answer differs. Who is taught what is determined by the consequences of choices about curricular goals, about how to handle student diversity, about how to use limited school resources, as well as many other factors. The overall result is that, for many students, access to educational possibilities in mathematics and the sciences is far more constrained than our cherished ideology would have us believe.

Access to various mathematical and science contents, according to the TIMSS data, varied considerably across the US and even within schools in the same (state or local) educational system, especially for eighth grade students. This occurred most often in mathematics in the form of tracking and in science in the form of participation in specialized courses (which might also reflect tracking).

Differential access seems to have resulted both from intentional and incidental factors. Variation within US fragmentation presented incidental differences in educational possibilities that resulted simply from variety in local educational decisions. However, *de facto* and even explicit tracking also characterized US schools and classrooms, especially in mathematics. In mathematics, at least, this seems to have represented intentional efforts to provide (presumably) 'more appropriate', differentiated access to courses, contents, and instructional activities.

One purpose of such differential access may have been to produce 'local homogeneity.' That is, sorting students into groups based on prior mathematical experience (with underlying assumptions of differing mathematical abilities) produces classes in which students are more likely to have roughly similar (homogeneous) attainments and readiness. This sorting would presumably be aimed at making instruction more effective for the individuals involved, as well as the groups containing them. This need for local homogeneity seems to have arisen, at least in part, from the fragmentation described earlier. This is true especially in a context of a highly mobile society in which individual students over a period of years may be exposed to some of the considerable variation in US mathematics and science education.

In other TIMSS countries, curricular goals and intentions included virtually no content differentiation through eighth grade in either mathematics or science. This kind of uniformity seems clearly to have been a matter of policy rather than merely incidental.

One TIMSS questionnaire was directed at principals, headmasters, etc. They were asked, among other questions, how many courses of study there were in their schools. For the most part, their answers confirmed the official policies of having one course of study at a given grade in mathematics or science. However, there were some countries (for example, Switzerland and Germany) for which the response to this question revealed that there was some content differentiation, even in spite of policies to the contrary.

However, this differentiation typically affected only a small percentage of students. In mathematics this typically involved only one alternative course type (see Exhibit 2.2 on page 18 or Exhibit 2.3 on page 20). In science for thirteen-year-olds, such alternate courses were essentially non-existent (see Exhibit 2.4 on page 21). This virtual absence of different course types was even more true for nine-year-olds, in mathematics as well as science.

Mathematics, as taught in US schools for eighth grade students, was in marked contrast to the rest of the TIMSS countries (again see Exhibit 2.2 and Exhibit 2.3). Those schools had as many as six different courses or tracks. Further, about 75 percent of US eighth graders were in schools that had more than one course type.

In mathematics, with its assumed incremental development, course types seemed, from their titles, likely to be based on where students were in their mastery of mathematics (for example, ready for a 'pre-algebra' course but not an 'algebra' course). Unfortunately, research on tracking in US schools suggests that curricular differences created in this way are typically (1) not limited to a single subject and (2) persist over time. Students are not placed into a single course type, but into a 'track' or 'course of study' in a broader sense that may involve more than one year and subjects other than mathematics.[5]

In short, US students may be assigned to 'general mathematics', 'pre-algebra', or 'algebra' on the basis of distinctions made in previous years, or even on the basis of distinctions made in areas other than mathematics (especially in smaller schools with limited course offerings so that students are 'typed' and an appropriate profile of courses, rather than a single course option, made available to them). Further, the distinctions introduced by these mathematics course types may persist into later grades and even into other courses which will conveniently fit the students' schedules. Therefore, students can be placed on different 'roads' that will continue to diverge and bring different experiences over the remainder of their schooling experience.

There seems to be little evidence of tracking in eighth grade science, other than that which occurred as an effect of mathematics tracking. Rather, science course types were differentiated by their focus on a specific discipline (earth, life, or physical science) or a melange of disciplines typical of earlier grades (general science). There seems to have been no assumption of incremental development of the sciences over the grades (except, perhaps, for development of science processes and methods). Within any educational system or its schools, the focus of science likely varied among grades. In each case, however, the result was still to differentiate content and thus to differentiate access to educational possibilities. A student's possibilities were shaped by the specific course taken. In both seventh and eighth grades about one-third of US students took 'general science' rather than a more focused discipline-oriented course.

This portrait of science course types is built around the TIMSS data on course titles and textbooks. We previously pointed out that the data on 'taught' science curricula in US education systems made it clear that the differences among the course types were more of degree rather than kind. Teacher coverage in discipline-oriented courses at the eighth grade level focused somewhat on topics from the putative central discipline, but, in every case, also included considerable instructional time devoted to topics from other disciplines. Thus, the clear distinctions between science course types seen in official curricula, titles, and textbooks were blurred by teacher topic coverage in 'taught' curricula.

WHO IS TAUGHT WHAT: A SUMMARY

- Access to educational possibilities, especially in the seventh and eighth grades, was greatly influenced in the US by the persistent practice of tracking in mathematics. Different course types were offered and students were given access to different profiles of contents covered, emphasized, or omitted according to the course type in which they were placed. As a part of this, however, constant review and content redundancy was introduced into 'taught' and official curricula.

- Tracking seems not to have been an important issue in science, even at the eighth grade. Science course types were differentiated by discipline-orientation. In the aggregate, this did not bode well for US eighth grade science achievement since most students had recently been exposed only to some of the broadly tested content. Fourth grade science, which did not yet show such specialization, seemed to be related to better overall achievement (as it was) since most students had been exposed to a wider sample of the tested content.

CONCLUDING REMARKS: WHAT WE TEACH, WHO WE TEACH

Differences among (local and state) education systems and the characteristic US practice of offering different educational possibilities to different student groups were reflected in what US teachers taught and to whom they taught it. This was seen in different amounts of time devoted to mathematics and science. There was considerable variation among teachers within the US on this instructional time but, even with this variation, the overall time was greater than in most other countries. US 'taught' curricula in science and especially mathematics, particularly in the middle grades, were highly redundant and assumed extensive review of contents to which students had been exposed in prior grades. This may have been because of perceived student mobility and providing different course types. This redundancy would have offset the benefits of more instructional time.

At the seventh and eighth grade levels, US curricula typically differentiated which contents were intended for different types of students. Different course types were offered, although the basis for distinctions differed for science and mathematics. The courses were related to different curricular emphases and different textbooks that reflected and enabled those emphases. This created differential access to possibilities for learning specific science and mathematics contents. In mathematics, this differentiation appears to have resulted in explicit and implicit tracking to provide for different levels of prior mastery and to take into account the assumption that mathematics content mastery was incremental and cumulative. Course types in science had a different basis and made no assumption of incremental learning. They were instead discipline-based. While having somewhat less serious consequences, these course types still produced differential access to educational possibilities.

The distinctions found in official curricula and textbooks for seventh and eighth grade students were blurred somewhat by teacher choices of what to cover, emphasize and omit. These extended the breadth of even discipline-based science courses. They resulted in less marked differences among mathematics course types (for example, 'pre-algebra' and 'general mathematics') at this level. Teacher perceptions added to the tendency for redundancy and extensive review already present in official mathematics curricula and textbooks. Given limited instructional time, this was also related to the shallowness of content coverage in mathematics.

Notes–

[1] The components of the exhibit are examples of what are called 'boxplots.' The 'box' encloses the middle 50 percent of values, going from the twenty-fifth to seventy-fifth percentile. The line in the box indicates the median values. The 'whiskers' indicate the middle 90 percent of values, going from the fifth to the ninety-fifth percentile.

[2] TIMSS used specially designed frameworks to characterize subject matter, whether this was in test items, textbook segments, official documents, or survey questionnaires. There was one framework for mathematics covering all the years of schooling and a comparable one for science. These documents provided a common language and a common category system for examining subject matter. Several aspects of subject matter were captured in these frameworks but perhaps the most essential was careful specification of specific subject matter content topics in mathematics and in the sciences. Framework development began with fairly encyclopedic listing of topics that were then reviewed by each of the TIMSS countries. The lists were modified but were considered too lengthy for most practical purposes. They were rearranged into 'nested' hierarchies of categories and sub-categories. Many individual items were combined to produce broader categories, even at the lowest, most specific levels. As a result, the mathematics framework had about 40 topics in ten major categories and science had about 80 topics in a slightly smaller number of categories.

[3] Recall that only about half as many topics were available for characterizing mathematics as were available for science.

[4] The TIMSS sampling design was based on the number of students in each grade. A stratified random sample of schools was selected with probability proportional to the number of students in the sampled grade. Mathematics classrooms were then randomly selected from each school and these students' teachers responded to the TIMSS Teacher Questionnaire. Responses from the Teacher and School questionnaires were weighted by the number of students represented. Consequently, although we refer to teacher and school characteristics, it is more accurate to interpret these in terms of students' exposure to the particular characteristics being considered. For example, as on page 38, the criterion of 'topics covered by 90 percent of teachers' means that 90 percent of the students had teachers who taught these topics.

[5] See Oakes, J. (1990). *Multiplying Inequalities: The Effects of Race, Social Class, and Tracking on Opportunities to Learn Mathematics and Science.* Santa Monica, CA: The RAND Corporation.

Chapter 3
How We Teach

At first glance, patterns of instructional activities seem to be the result of choices and decisions made by individual teachers. In part, they are. However, if strong patterns of common behavior characterize large groups of teachers in similar situations (for example, eighth grade science teachers), this is likely evidence of common factors shaping those instructional practices.

These common factors may come from the discipline involved (for example, the nature of science as presented to students at a certain stage of development). The common factors may also stem from the specific educational systems that (1) trained the teachers; (2) gave them their missions, resources, and working climate; or (3) helped shape their experiences as teachers. Even teachers' conceptions of the subject matter discipline appropriate for students at a specific age or grade (for example, eighth grade mathematics), and how this should be communicated, likely flow from their training and experiences – from their professional life within specific US educational systems.

CHARACTERIZING INSTRUCTIONAL PRACTICES

Are there patterns in the instructional practices of US science and mathematics teachers? Are the characteristic patterns for US teachers that different from those of other countries' teachers? Are there differences between instructional practices of US science teachers compared to US mathematics teachers? Answering these and similar questions requires finding ways to characterize instructional practices.

For this report, we used four approaches to characterize instructional practices through the TIMSS data. First, we described specific practices or teacher behaviors. No one study could gather data relevant to more than a limited set of features and behaviors. How did we choose the features on which we would gather data and the means by which those data would be gathered? A pilot project, the Survey of Science and Mathematics Opportunity (hereafter SMSO) informed our decisions.[1] In SMSO, six countries (France, Japan, Norway, Spain, Switzerland, and the United States) worked together in instrument development for TIMSS. The multinational SMSO team gathered preliminary data through classroom observations, teacher logs and interviews, and by overseeing completion of preliminary survey questions.

As a result, the TIMSS surveys of instructional practices and teacher behaviors were designed to capture characteristic patterns and cross-national differences if these existed – as the SMSO development and TIMSS pilot data suggested they did. Only a sample of these data can be reported here – in particular, those most relevant to portraying the similarities and differences of US science and mathematics teachers compared to their counterparts in other TIMSS countries.

Second, we formed typologies – sets of types or categories – of teachers based on their reports of one of their own typical lesson's structure. The mix of more global activities (review, seatwork, etc.) from which mathematics and science lessons could be structured emerged during SMSO development. Questions were asked on TIMSS surveys to capture the use of these global activities in typical lessons. Empirical, statistical methods were used to identify the categories (types of teachers that differed in important ways in their use of these global activities).

Third, we formed typologies of teachers' beliefs about their subject matter disciplines and pedagogical approaches. In this case, it was the teachers' beliefs and opinions that contributed to patterns in teacher instructional practices. For that reason, our discussion of data from these belief typologies is found in the next chapter so that the current chapter can be restricted to behaviors and practices.

Fourth, since most data described thus far relied on teachers' reports of their own practices, behaviors, and beliefs, we sought a source to 'triangulate' with teacher self-report survey data to help assess the validity of those data. We developed and used questionnaires to gather information from students on several aspects of their beliefs, behaviors, and experiences. Among these were their perceptions of their own teachers' typical practices in teaching mathematics and science.

These four methods combine to provide an empirical basis for the search for characteristic instructional practices and teacher beliefs. These data should reveal and document such group practices if they exist. The remainder of this chapter investigates these data, except for those on teacher beliefs which are postponed to the following chapter.

US SCIENCE AND MATHEMATICS TEACHERS' LESSON STRUCTURES

Lesson 'Building Blocks.' Our investigation of the 'building blocks' of US mathematics and science teachers' lesson structures focuses on five global behaviors: reviewing the content covered in a previous lesson, reviewing homework assigned in a previous lesson, providing instruction on new subject matter, having students work on in-class exercises that were either used in lesson development or otherwise discussed in the lesson, and having students work on 'homework' that would not be discussed until a later lesson. These five activities were found to characterize segments of typical lessons for many teachers and to reveal differences among patterns for teachers from different countries.

These global activities are quite generic. Most teachers do these – review, work with homework, provide instruction on new content, etc. – to some extent. What was surprising was the extent to which certain groups of teachers used these activities. To document 'extent', we established empirical 'cut-points' that separated teachers who pursued a particular activity to an unusual degree from those who pursued it to a more common degree. The cut-points differed for each of the five activities.

Exhibit 3.1 on page 66 exhibits the percentage of fourth and eighth grade mathematics teachers at or above the selected cut-points for the five lesson 'building block' activities discussed above. About half of all eighth grade teachers in both mathematics and science (see Exhibit 3.2) spent at least five minutes reviewing content from the previous lesson. The variations in the use of the other four activities were more pronounced.

From these data it is clear that reviewing homework for more than 10 minutes was quite common among US eighth grade mathematics teachers, and was practiced by about 29 percent of them. This contrasts, as the exhibit shows, with Japan, the Russian Federation, the Czech Republic and Korea, among others. This was also more than was done by US fourth grade mathematics teachers (about 17 percent) and eighth grade science teachers (about 11 percent). It seems to have been a common characteristic of US middle grade mathematics.

About 40 percent of both US fourth and eighth grade mathematics teachers provided instruction on new content for at least 20 minutes during a lesson. This was also true for US science teachers (see Exhibit 3.2). About 60 percent of Japanese eighth grade teachers reached this cut-point as did 64 percent of Japanese eighth grade science teachers. US eighth grade science teachers reached this 20 minute criterion less often than did the teachers from many of the other countries examined.

US teachers also more often spent over 10 minutes (of a typically 50 minute lesson) reviewing homework from a previous lesson. Over one fourth of US eighth grade mathematics teachers did this, as did their counterparts in France. This was also true for more than half the teachers in several countries including Spain, the Netherlands and Israel. In eighth grade science, the percentages doing this were far smaller, but still more pronounced, for the US than for other countries except again, Israel, the Netherlands, and Spain (see Exhibit 3.2). This seems to be clear evidence that mathematics, unlike science, in the

Exhibit 3.1. *Percentage of Fourth and Eighth Grade* Mathematics Teachers** at or Above Select Cut-Points for Five Lesson Building Blocks.*

Grade 4	Review Lesson 5 Minutes or More	Review Homework 10 Minutes or More	Instruction 20 Minutes or More	Exercises 15 Minutes or More	Do Homework 15 Minutes or More
USA	42.4 (3.1)	16.8 (2.7)	38.7 (3.8)	16.4 (3.5)	20.5 (4.0)
Australia	50.8 (4.3)	3.1 (1.3)	18.0 (2.9)	43.0 (4.2)	1.1 (0.9)
Canada	56.6 (4.8)	13.0 (2.3)	20.5 (3.4)	41.1 (4.8)	14.8 (2.4)
Czech Republic	81.4 (3.5)	3.1 (1.4)	13.7 (3.0)	41.0 (4.3)	0.1 (0.1)
Hong Kong	22.6 (4.3)	3.4 (1.5)	51.9 (5.1)	7.7 (2.4)	5.5 (2.7)
Hungary	50.2 (4.5)	7.1 (2.2)	25.6 (4.2)	33.6 (4.1)	0.9 (0.9)
Israel	55.7 (8.6)	29.8 (7.9)	20.0 (6.8)	18.1 (6.4)	2.7 (0.4)
Japan	50.8 (3.7)	11.2 (2.3)	58.7 (4.5)	34.5 (4.2)	3.0 (1.4)
Korea	39.8 (3.6)	8.2 (2.5)	17.6 (3.1)	17.8 (3.2)	28.3 (3.5)
Netherlands	23.1 (4.6)	7.3 (3.3)	28.2 (5.3)	75.6 (5.5)	0.8 (0.8)
New Zealand	53.4 (4.9)	4.0 (1.7)	24.5 (4.0)	34.3 (4.2)	0.4 (0.0)
Norway	45.5 (4.4)	12.6 (3.5)	10.9 (3.4)	66.3 (4.4)	0.4 (0.0)
Singapore	36.3 (3.8)	12.4 (2.7)	42.5 (3.7)	24.2 (3.6)	12.6 (2.6)
Thailand	77.5 (4.2)	46.6 (6.2)	79.7 (4.1)	65.6 (5.4)	15.1 (3.4)

Grade 8	Review Lesson 5 Minutes or More	Review Homework 10 Minutes or More	Instruction 20 Minutes or More	Exercises 15 Minutes or More	Do Homework 10 Minutes or More
USA	45.0 (3.8)	28.5 (3.7)	40.4 (4.1)	11.9 (2.0)	31.6 (4.4)
Australia	46.8 (3.2)	8.1 (1.5)	19.3 (2.7)	60.7 (3.4)	2.0 (1.0)
Belgium (Fl)	43.7 (5.3)	14.4 (3.0)	29.9 (4.0)	48.8 (4.3)	3.2 (2.5)
Canada	43.3 (4.5)	25.5 (3.0)	22.9 (4.2)	15.0 (3.0)	31.4 (3.7)
Czech Republic	78.0 (4.3)	3.3 (1.7)	17.8 (4.4)	55.5 (5.7)	0.8 (0.8)
France	22.2 (3.9)	38.5 (4.6)	35.8 (4.8)	25.2 (4.7)	9.0 (2.2)
Germany	58.5 (5.1)	23.4 (4.6)	37.4 (5.5)	25.2 (4.7)	0.0
Hong Kong	48.5 (6.1)	16.6 (3.5)	48.0 (5.8)	31.0 (5.8)	14.3 (3.4)
Hungary	54.4 (3.6)	23.5 (3.6)	33.1 (3.9)	31.6 (3.7)	0.0
Israel	66.5 (7.9)	59.2 (6.8)	15.8 (6.1)	23.4 (7.6)	0.0
Japan	67.3 (3.8)	18.2 (3.8)	60.2 (4.2)	40.2 (4.2)	2.4 (1.4)
Korea	47.8 (4.5)	10.4 (3.1)	53.6 (4.1)	15.9 (3.4)	1.4 (1.0)
Netherlands	39.9 (6.0)	78.7 (4.5)	7.2 (3.1)	19.6 (4.5)	68.5 (5.3)
New Zealand	33.5 (4.3)	3.4 (1.5)	14.5 (3.0)	64.5 (4.2)	0.8
Norway	52.2 (5.5)	20.7 (4.3)	15.4 (3.5)	61.2 (5.3)	6.4 (2.6)
Russian Federation	55.1 (4.9)	5.5 (2.5)	19.5 (3.6)	10.1 (4.2)	0.0
Singapore	33.3 (4.8)	10.8 (3.2)	66.4 (4.9)	14.2 (3.4)	19.7 (4.1)
Spain	38.9 (4.9)	53.0 (4.5)	37.1 (4.7)	24.1 (4.2)	8.8 (2.8)
Sweden	46.3 (3.7)	6.0 (2.0)	10.6 (2.3)	77.0 (3.1)	3.8 (1.5)
Switzerland	43.0 (5.2)	18.2 (3.4)	32.9 (4.9)	28.6 (4.6)	13.0 (2.8)
Thailand	86.0 (4.5)	39.1 (6.6)	36.3 (5.8)	25.3 (6.5)	30.2 (5.9)

* See note 3 on page 12.
** See note 4 on page 61.

US was built around incremental development with an emphasis, not just on a brief review of previous lesson content, but also a more extended discussion of work from the previous lesson.

Far more Japanese eighth grade mathematics teachers used in-class exercises than did their US counterparts. Other evidence[2] suggests that there were qualitative differences in how this student work was used in developing new

Exhibit 3.2. *Percent of Eighth Grade Science Teachers at or Above Select Cut-Points for Five Lesson Building Blocks.*

	Review Lesson 5 Minutes or More		Review Homework 10 Minutes or More		Instruction 20 Minutes or More		Exercises 15 Minutes or More		Do Homework 10 Minutes or More	
USA	51.7	(5.1)	11.3	(2.8)	43.5	(4.8)	7.6	(2.0)	11.9	(2.8)
Australia	44.8	(4.3)	7.6	(1.9)	28.3	(3.7)	26.1	(3.4)	3.8	(1.1)
Belgium (Fl)	42.4	(4.6)	15.2	(3.6)	56.6	(5.7)	7.0	(2.2)	0.8	(0.8)
Canada	33.2	(3.7)	12.4	(2.9)	29.4	(3.9)	11.7	(4.0)	7.4	(2.7)
Czech Republic	80.6	(2.9)	0.4	(0.4)	68.0	(2.8)	4.1	(1.2)	1.1	(1.0)
France	22.2	(2.9)	8.1	(1.7)	52.9	(3.1)	10.1	(2.1)	0.0	
Germany	64.5	(4.3)	5.5	(1.9)	61.9	(4.9)	9.2	(2.5)	1.1	(1.0)
Hong Kong	41.9	(6.1)	4.0	(2.3)	57.9	(5.9)	10.6	(3.7)	7.4	(3.1)
Hungary	53.1	(2.6)	6.1	(1.4)	59.9	(2.4)	9.2	(1.3)	0.3	(0.3)
Israel	52.5	(9.7)	23.1	(6.8)	52.0	(8.1)	9.5	(5.0)	3.5	(0.3)
Japan	61.7	(4.0)	1.9	(1.3)	64.4	(4.4)	10.6	(2.9)	2.5	(1.5)
Korea	36.4	(4.2)	7.9	(2.8)	61.9	(4.0)	3.8	(1.5)	0.0	
Netherlands	47.8	(3.8)	61.7	(4.1)	32.3	(2.8)	17.5	(2.7)	24.4	(3.4)
New Zealand	28.7	(3.5)	3.2	(1.5)	18.2	(3.3)	22.7	(3.2)	0.0	
Norway	41.5	(5.4)	10.1	(2.8)	54.4	(5.5)	22.9	(4.6)	5.1	(2.3)
Russian Federation	75.6	(2.6)	2.6	(0.8)	48.8	(2.7)	9.1	(1.7)	0.7	(0.5)
Singapore	34.2	(4.4)	5.3	(1.8)	79.7	(3.7)	3.5	(1.7)	11.4	(3.3)
Spain	31.4	(4.4)	34.8	(4.6)	43.9	(4.9)	16.6	(3.7)	7.4	(2.6)
Sweden	32.5	(3.9)	3.7	(1.7)	24.4	(3.4)	8.8	(2.1)	4.0	(1.5)
Switzerland	44.1	(5.2)	1.7	(1.3)	60.8	(5.4)	11.7	(3.5)	0.8	(0.1)
Thailand	55.1	(6.9)	12.1	(3.7)	33.1	(7.5)	10.4	(4.0)	14.1	(5.3)

lesson content as well as quantitative differences in how much time was devoted to it. Such differences were not pronounced in eighth grade science, nor was this use of 'exercises' in science classes as common in any country examined as it was in mathematics. The use of exercises, in one way or another, appears to have been a characteristic feature of mathematics instruction.

There was one other difference in the use of these 'building blocks' in which US teachers stood out. Many US eighth grade mathematics teachers (more than one fourth) spent 10 minutes or more of in-class 'instructional' time allowing students to work on homework related to the day's lesson, although that work would not be discussed or evaluated until the next lesson. A smaller number of eighth grade science teachers used this method. The percentage of eighth grade science or mathematics teachers doing this was negligible in the other countries investigated except for Canada, Thailand and the Netherlands. Comparatively extensive use of class time for 'homework' (placed here in quotation marks since the context makes the meaning of the term ambiguous) seems to be a characteristic US instructional practice, especially in mathematics. This might indicate 'slack time' created by redundant middle school mathematics curricula. Alternatively, it might indicate low motivation among US students to pursue true homework that made it necessary to have them begin the 'homework' under supervision.

Lesson Structure Typology. As described earlier, we also examined lesson structure by constructing empirical category systems (typologies) into which teachers could be 'placed' based on their instructional practices. (Still other typologies to be discussed in the next chapter examined teacher beliefs and opinions relevant to the nature of school mathematics and science). We used the same global activities (review, etc.) as a basis for this lesson structure typology, with two distinctions. First, since 'review of previous lesson's material' alone did not distinguish well among teachers and countries, it was combined with 'homework review' to form a single 'review of old content' category. Second, in-class work, which was done during the course of a lesson *and* used as part of that lesson's development of subject matter, was labeled by the traditional US educational term 'seatwork.' In-class work based on the lesson's content, but to be discussed (if at all) in future lessons, was labeled by the traditional term 'homework', even though 'in-class homework' is an oxymoron. These two were combined to form the single category which we termed seatwork.

We examined which activities dominated teachers' use of lesson time. Teachers were clustered into groups in terms of which activity or activities dominated their typical lesson's instruction. A label was given for each group of teachers in terms of which one or two activities dominated their instructional practice during a typical lesson. Seven categories accounted for most teachers in each subject discipline, grade level, and country. The remainder were put into a catch-all 'other' category.

Exhibit 3.3 on page 69 shows the percentage of fourth grade teachers[3] in each of the seven categories and 'other.' The dominant category for US fourth grade teachers in mathematics was that for those whose instruction was devoted primarily to 'seatwork.' About one-fifth of these teachers used seatwork as their dominant instructional activity. Typically 'seatwork' was done by individual students in their own seats (hence the traditional name) rather than in small groups. During this time the teacher might circulate to provide help to individual students, or simply be available as a resource to which students could go for help. In other cases, seatwork was for a limited period of time and its results integrated into further development of the day's lesson.[4]

In eighth grade mathematics (see Exhibit 3.4), the dominant US patterns were 'seatwork combined with homework review' (24 percent of teachers) and 'seatwork' (almost 20 percent of teachers) closely followed by 'instruction on new material combined with seatwork' (almost 16 percent of teachers). Thus these three seatwork-oriented patterns accounted for over one-half of the US eighth grade mathematics' teachers dominant activities. This reliance on seatwork by mathematics teachers of thirteen-year olds was fairly common across many TIMSS countries.

In contrast, the dominant pattern for Japanese eighth grade mathematics teachers was 'new instruction combined with seatwork' (27 percent of teachers).

Exhibit 3.3. *Percent of Fourth Grade Teachers in Each Lesson Structure Category.*

	Review	Seatwork	Review & Seatwork	Instruction	Review & Instruction	Instruction & Seatwork	Review, Instruction, Seatwork	Other
USA	11.2 (2.9)	19.3 (3.3)	18.0 (2.4)	7.5 (1.8)	2.8 (1.6)	14.8 (3.1)	13.0 (2.9)	13.3 (2.5)
Australia	1.5 (0.8)	40.6 (3.9)	14.7 (2.6)	7.3 (2.0)	0.5 (0.3)	9.2 (2.2)	1.1 (0.7)	25.2 (3.1)
Canada	8.3 (2.2)	28.8 (4.0)	22.0 (2.9)	2.9 (1.3)	2.9 (1.4)	7.8 (2.2)	6.8 (2.4)	20.4 (5.7)
Czech Republic	12.4 (2.7)	27.8 (3.8)	41.1 (4.4)	4.8 (1.8)	3.8 (1.6)	2.8 (1.4)	2.3 (1.3)	5.0 (1.6)
Hong Kong	4.4 (1.8)	18.5 (4.3)	15.2 (3.8)	20.7 (4.5)	3.9 (1.5)	26.0 (4.1)	1.2 (0.8)	9.9 (3.5)
Hungary	9.7 (2.9)	27.6 (4.1)	26.0 (4.1)	8.7 (2.4)	10.0 (2.8)	4.3 (1.7)	2.6 (1.4)	11.1 (3.1)
Israel	15.7 (6.4)	14.0 (5.6)	26.9 (7.2)	2.9 (2.9)	11.7 (5.5)	3.0 (2.9)	2.5 (2.5)	23.4 (7.5)
Japan	2.6 (1.5)	16.9 (3.1)	10.7 (2.6)	15.1 (3.1)	3.6 (1.7)	32.2 (4.5)	7.8 (2.5)	11.0 (2.6)
Korea	5.1 (1.9)	38.6 (4.1)	19.2 (3.5)	3.5 (1.6)	0.5 (0.5)	6.1 (1.8)	7.6 (2.3)	19.5 (3.5)
Netherlands	3.5 (2.4)	53.1 (6.6)	8.7 (3.5)	1.9 (1.8)	1.5 (1.6)	23.6 (4.9)	1.1 (1.1)	6.6 (3.0)
New Zealand	8.1 (2.5)	36.4 (4.2)	10.9 (3.1)	12.3 (2.7)	4.3 (2.0)	6.5 (2.4)	1.4 (0.9)	20.0 (3.6)
Norway	5.1 (2.2)	42.8 (4.5)	26.3 (4.9)	5.9 (2.4)	0.0	5.0 (2.4)	0.0	14.9 (3.5)
Singapore	9.6 (2.5)	22.0 (3.2)	16.2 (2.5)	13.4 (2.8)	8.7 (2.6)	11.9 (2.6)	8.5 (2.5)	9.7 (2.3)
Thailand	5.4 (2.0)	3.7 (2.1)	5.6 (2.2)	4.4 (2.0)	7.9 (3.7)	2.4 (1.1)	64.9 (6.1)	5.7 (1.9)

Exhibit 3.4. *Percent of Eighth Grade Mathematics Teachers in Each Lesson Structure Category.*

	Review	Instruction	Seatwork	Review & Instruction	Review & Seatwork	Instruction & Seatwork	Review, Instruction, Seatwork	Other
USA	9.6 (2.2)	9.6 (3.7)	19.9 (2.7)	4.0 (1.6)	24.0 (3.7)	16.1 (3.3)	10.7 (2.2)	6.1 (1.8)
Australia	3.8 (1.4)	4.8 (1.5)	46.4 (3.0)	0.9 (0.6)	23.6 (2.7)	9.9 (1.8)	3.8 (1.5)	6.9 (1.8)
Belgium (Fl)	3.7 (1.7)	8.0 (2.2)	21.2 (3.9)	2.8 (1.4)	18.5 (4.1)	11.8 (2.8)	7.3 (3.0)	26.7 (3.9)
Canada	13.0 (4.2)	3.1 (2.3)	22.1 (3.1)	3.9 (2.3)	28.5 (3.8)	6.5 (2.0)	9.4 (2.4)	13.5 (3.3)
Czech Republic	7.0 (2.7)	5.3 (2.5)	44.1 (4.5)	1.0 (1.0)	26.4 (5.2)	6.8 (3.3)	4.6 (1.6)	4.7 (1.9)
France	6.6 (2.4)	3.7 (1.8)	17.3 (3.9)	10.6 (3.1)	21.0 (4.5)	10.7 (3.3)	10.8 (3.3)	19.2 (4.3)
Germany	11.7 (3.2)	5.5 (2.8)	13.1 (3.3)	12.9 (4.3)	23.2 (4.5)	11.3 (3.8)	7.8 (2.9)	14.6 (3.8)
Hong Kong	2.7 (1.9)	6.6 (3.0)	25.0 (5.2)	8.1 (3.7)	20.2 (4.6)	22.0 (4.7)	11.4 (3.9)	4.1 (2.4)
Hungary	8.0 (2.2)	6.6 (2.2)	16.9 (3.4)	12.5 (3.0)	36.8 (4.0)	5.9 (2.1)	8.1 (2.4)	5.2 (1.9)
Israel	16.2 (5.9)	6.5 (3.7)	4.4 (3.4)	2.9 (2.9)	53.9 (10.1)	1.1 (1.1)	5.3 (4.1)	9.7 (4.2)
Japan	2.4 (1.2)	8.2 (2.5)	17.3 (3.5)	4.6 (1.8)	14.8 (3.7)	27.0 (3.9)	20.4 (3.6)	5.3 (1.5)
Korea	4.6 (1.7)	11.8 (3.1)	17.5 (3.1)	5.6 (2.0)	15.5 (3.4)	24.8 (3.4)	11.4 (3.2)	8.9 (2.6)
Netherlands	6.7 (3.0)	0.0	11.4 (3.6)	0.9 (0.9)	73.4 (4.7)	0.6 (0.6)	5.8 (2.9)	1.2 (1.2)
New Zealand	2.2 (1.3)	6.9 (2.2)	56.7 (4.2)	0.5 (0.5)	17.6 (3.0)	6.2 (2.1)	0.8 (0.8)	9.0 (2.4)
Norway	10.6 (3.1)	1.5 (1.0)	30.2 (4.6)	1.0 (1.0)	33.7 (5.0)	10.3 (3.1)	2.6 (1.6)	10.1 (3.1)
Russian Federation	22.3 (3.9)	6.7 (2.1)	29.8 (5.2)	2.6 (1.4)	18.2 (3.4)	7.2 (2.4)	2.9 (1.2)	10.2 (2.9)
Singapore	2.5 (1.3)	17.8 (4.0)	17.6 (3.7)	1.8 (1.1)	9.7 (3.1)	35.0 (3.9)	11.7 (2.9)	3.7 (1.8)
Spain	12.5 (3.4)	1.7 (1.2)	7.1 (2.4)	9.8 (3.0)	35.4 (4.8)	14.9 (3.8)	10.7 (3.4)	7.8 (2.7)
Sweden	1.7 (1.0)	3.5 (1.5)	61.9 (3.5)	0.0	14.5 (2.8)	5.1 (1.7)	2.0 (1.2)	11.3 (2.3)
Switzerland	11.5 (3.0)	7.8 (2.6)	21.4 (3.9)	4.6 (1.8)	23.5 (4.3)	13.0 (3.7)	7.5 (2.9)	10.7 (3.1)
Thailand	9.9 (3.8)	3.1 (2.4)	5.9 (2.0)	6.1 (2.9)	41.2 (6.3)	2.6 (1.3)	24.6 (4.8)	6.6 (2.9)

Far fewer US teachers used this pattern (about 15 percent in mathematics at both fourth and eighth grades). As discussed earlier, there is evidence of qualitatively different use of seatwork by Japanese teachers. Further, they tended more often to combine seatwork with instruction on new material, while US eighth grade mathematics teachers used seatwork alone or combined it with review of previous material. This pattern in eighth grade mathematics can be somewhat directly validated by the complementary videotape study of eighth grade mathematics classes in Japan, Germany, and the US.[5]

Exhibit 3.5. *Percent of Eighth Grade Science Teachers in Each Lesson Structure Category.*

	Review	Instruction	Seatwork	Review & Instruction	Review & Seatwork	Instruction & Seatwork	Review, Instruction, Seatwork	Other
USA	15.6 (4.1)	16.0 (3.1)	8.7 (2.2)	11.9 (2.9)	10.2 (3.2)	12.9 (3.6)	2.7 (1.0)	22.0 (3.9)
Australia	5.8 (1.7)	11.7 (2.4)	21.3 (3.8)	3.7 (1.8)	12.6 (3.1)	10.8 (2.8)	2.1 (1.2)	31.9 (3.8)
Belgium (Fl)	3.3 (1.4)	23.5 (3.8)	3.2 (1.9)	15.8 (3.6)	3.5 (1.6)	14.2 (5.4)	3.2 (1.7)	33.3 (4.6)
Canada	6.6 (1.5)	11.8 (3.3)	21.5 (4.8)	7.0 (2.7)	8.6 (1.8)	6.3 (2.1)	4.3 (1.9)	33.9 (5.1)
Czech Republic	8.1 (1.9)	25.0 (2.6)	8.0 (1.8)	23.6 (2.5)	4.9 (1.3)	12.1 (2.1)	7.2 (1.7)	11.1 (2.2)
France	3.4 (1.2)	26.9 (2.7)	9.7 (2.3)	8.2 (2.0)	2.8 (1.2)	13.6 (2.7)	4.3 (1.5)	31.2 (3.6)
Germany	7.0 (2.2)	26.3 (4.6)	14.2 (3.0)	12.9 (2.9)	5.5 (2.0)	20.4 (3.9)	2.4 (1.4)	11.3 (2.8)
Hong Kong	4.8 (2.8)	30.9 (5.6)	6.9 (3.1)	3.0 (2.1)	2.2 (1.6)	16.2 (4.6)	7.9 (3.3)	28.2 (5.3)
Hungary	6.1 (1.0)	22.6 (2.3)	10.7 (1.7)	15.5 (2.3)	8.9 (1.5)	17.2 (2.0)	4.6 (1.1)	14.5 (1.8)
Israel	1.4 (1.4)	4.3 (3.4)	20.1 (7.5)	15.6 (6.1)	9.4 (5.1)	13.9 (5.6)	18.1 (6.2)	17.1 (6.8)
Japan	2.4 (1.4)	34.0 (4.7)	7.4 (2.1)	3.6 (1.7)	3.1 (1.6)	25.2 (3.8)	1.6 (1.2)	22.7 (3.4)
Korea	2.6 (0.9)	36.4 (3.9)	8.5 (2.1)	14.0 (3.2)	1.0 (0.8)	8.9 (2.2)	2.6 (1.2)	25.9 (3.7)
Netherlands	16.3 (2.6)	4.2 (1.3)	7.6 (2.2)	8.9 (2.3)	39.1 (3.3)	8.6 (2.4)	10.5 (2.5)	4.7 (1.5)
New Zealand	7.0 (2.0)	5.3 (1.9)	41.3 (4.0)	4.2 (1.7)	3.9 (1.7)	5.8 (1.9)	2.8 (1.4)	29.7 (3.5)
Norway	7.9 (2.9)	29.1 (5.2)	16.2 (3.8)	6.2 (2.5)	4.5 (1.8)	15.7 (4.1)	3.4 (1.9)	17.1 (4.2)
Russian Federation	16.9 (2.5)	12.4 (2.1)	11.2 (1.7)	17.9 (2.5)	11.7 (2.4)	8.3 (1.7)	10.3 (2.7)	11.3 (1.8)
Singapore	0.0	35.9 (4.4)	5.9 (2.3)	7.1 (2.4)	5.3 (2.2)	33.3 (4.9)	3.4 (1.6)	9.1 (2.0)
Spain	12.3 (3.0)	7.3 (2.6)	7.0 (2.7)	7.5 (3.0)	25.0 (4.3)	14.6 (3.5)	14.5 (3.4)	11.9 (3.2)
Sweden	6.3 (2.1)	14.4 (2.5)	14.5 (2.3)	1.5 (0.8)	2.4 (1.0)	7.2 (2.2)	1.2 (0.4)	52.6 (3.8)
Switzerland	3.0 (2.1)	34.0 (4.9)	12.7 (3.6)	9.1 (3.0)	1.7 (1.5)	14.9 (3.3)	2.7 (1.8)	21.9 (4.0)
Thailand	5.9 (2.7)	12.4 (5.4)	26.0 (6.6)	5.4 (3.3)	10.8 (4.1)	5.3 (2.3)	10.0 (4.1)	24.2 (6.2)

The patterns were quite different for eighth grade science (see Exhibit 3.5 on page 70), perhaps reflecting the different places of discrete 'exercises' suitable for seatwork. In Japan, the dominant category was that of 'instruction on new material' (about 35 percent of the teachers). In the US, no one of the seven categories was markedly dominant, although almost 15 percent fell into the 'instruction on new material' category. The largest proportion (over 20 percent) of US teachers fell into the 'other' category, which was really an aggregate of several other patterns. Eighth grade science teachers were most often involved in groups that included instruction on new material as one of their dominant activities (about 40 percent combined for science compared to about 30 percent for mathematics).

US SCIENCE AND MATHEMATICS TEACHERS' LESSON STRUCTURES: A SUMMARY

We examined characteristic lesson structures for US mathematics and science teachers, investigating the balance of typical activities and what percent of teachers would fall into categories dominated by different primary activities. Several results emerged including the following:

- US mathematics instruction for both nine- and thirteen-year-olds seems to have been dominated by seatwork and reviewing homework. Teachers spent some time teaching new material, but it was not the dominant feature of lessons. Less than 40 percent of US teachers provided 20 minutes or more instruction on new material during a class period. Japanese teachers, by contrast, spent most of their time on a combination of instruction on new material and on seatwork which was, for the most part, actively tied to the instruction of new material during the course of a lesson.

- How US teachers used their lesson time in science instruction differed from that of mathematics instruction. In science, the dominant pattern was instruction on new material or instruction combined with seatwork. More than 40 percent of eighth grade science teachers provided 20 minutes or more of instruction on new material.

OTHER ASPECTS OF INSTRUCTIONAL PRACTICE

One source of 'triangulating' information in order to confirm and extend patterns already revealed by teacher self-report survey data was the data from the TIMSS student questionnaires. Among the many questions on the TIMSS student questionnaire were a number of questions that dealt with students' perceptions of their own instructional experiences in mathematics and science classrooms.

Characteristic Lesson Activities. Exhibit 3.6 on page 72 displays the percent of eighth grade mathematics students who chose 'always' or 'often' as a response to survey questions about what their teachers did during class instruction. These data make it clear that US mathematics students perceived their teachers as giving tests far more often than did the students in other countries, with the exception of students in Belgium (Flemish). US eighth graders also indicated far more often than some of their counterparts that their teachers asked them to work on their own. Over 85 percent of the students indicated that they always or often worked on their own in class, while only 40 percent or less of Japanese, Koreans, French and Germans indicated this. A high percentage of students in six other countries also indicated that they always or often worked on their own. This is consistent with the lesson structure typology which showed that US teachers tended mostly to fall into categories dominated by seatwork, while Japanese teachers fell more often into categories involving new instruction. Seatwork most typically involved students working on their own (although US students do indicate some pair and small group work). Further, about 75 percent of the US students (contrasted with about 25 percent or less of Japanese, Korean or Czech Republic students) indicated that they began their 'homework' in class.

Exhibit 3.6. Eighth Grade Students' Perceptions of Classroom Activities in Mathematics (percent responding 'always' or 'often').

	Teacher Shows Us How to Do Problems	Copy Notes From the Board	Have a Quiz or Test	Work on Our Own	Work on Projects	Use Calculators	Use Computers
USA	94.2 (0.6)	66.9 (1.8)	85.1 (0.9)	85.5 (0.7)	26.6 (1.2)	69.7 (2.6)	10.5 (1.5)
Australia	91.7 (0.5)	69.9 (1.4)	53.7 (1.2)	90.0 (0.7)	21.4 (1.3)	86.1 (1.7)	5.1 (0.9)
Belgium (Fl)	82.4 (1.2)	81.3 (1.4)	92.9 (0.8)	65.3 (1.8)	28.2 (1.1)	29.5 (3.8)	1.6 (0.6)
Canada	90.1 (0.8)	54.7 (2.2)	72.5 (1.3)	89.7 (0.6)	23.4 (1.2)	71.4 (2.1)	4.9 (0.4)
Czech Republic	85.1 (1.4)	73.0 (1.6)	28.2 (1.3)	54.5 (2.0)	4.9 (0.5)	61.2 (3.0)	3.6 (1.8)
England	89.6 (1.0)	64.3 (1.9)	49.6 (1.4)	89.7 (1.0)	37.3 (1.8)	90.8 (0.9)	9.3 (1.2)
France	79.7 (1.6)	84.5 (1.1)	70.5 (1.4)	36.5 (1.3)	26.1 (1.1)	70.3 (1.8)	3.8 (0.8)
Germany	71.1 (1.9)	70.1 (1.6)	34.3 (2.0)	41.5 (1.5)	14.8 (1.0)	55.7 (2.7)	4.8 (0.7)
Hong Kong	92.3 (1.0)	62.1 (2.8)	79.2 (2.2)	76.2 (1.3)	16.1 (1.4)	82.5 (2.8)	3.4 (0.4)
Hungary	76.3 (1.2)	74.5 (1.2)	20.1 (1.2)	66.0 (1.8)	51.7 (1.0)	40.5 (2.3)	2.4 (0.4)
Israel	88.4 (1.2)	77.8 (2.4)	56.6 (3.3)	67.8 (2.7)	11.4 (1.6)	71.6 (3.1)	11.3 (3.0)
Japan	87.6 (0.8)	87.8 (1.4)	41.3 (2.3)	36.9 (1.4)	8.3 (1.0)	3.9 (0.7)	4.3 (1.2)
Korea	86.6 (0.8)	59.9 (2.5)	26.3 (1.5)	25.9 (1.1)	69.9 (1.0)	2.0 (0.3)	1.9 (0.3)
Netherlands	66.7 (2.6)	36.4 (2.6)	54.9 (1.6)	83.0 (1.8)	2.2 (0.4)	90.3 (1.3)	1.7 (0.4)
New Zealand	89.9 (0.8)	83.6 (1.4)	55.0 (1.7)	88.6 (0.8)	27.8 (1.8)	73.2 (2.4)	4.3 (0.6)
Norway	84.5 (0.9)	44.4 (1.7)	34.0 (1.3)	50.3 (1.5)	7.6 (0.6)	71.6 (1.9)	1.9 (0.3)
Russian Federation	79.6 (1.3)	78.2 (1.5)	77.1 (1.5)	62.3 (1.5)	11.0 (0.8)	53.9 (2.8)	2.2 (0.3)
Singapore	96.6 (0.4)	56.4 (2.6)	72.8 (1.2)	67.0 (1.1)	5.4 (0.5)	82.4 (1.7)	1.8 (0.4)
Spain	89.9 (0.9)	58.1 (1.8)	75.5 (1.4)	67.6 (1.3)	24.6 (1.2)	28.6 (2.7)	3.0 (0.7)
Sweden	87.0 (0.9)	37.0 (1.7)	56.8 (1.6)	85.7 (0.7)	31.8 (1.0)	53.5 (2.5)	9.4 (1.1)
Switzerland	82.5 (1.0)	44.7 (2.1)	58.6 (1.2)	70.4 (1.3)	21.4 (1.1)	32.3 (2.1)	3.9 (0.6)
Thailand	90.9 (0.9)	85.1 (1.2)	58.8 (1.7)	51.5 (1.1)	12.4 (0.9)	7.1 (1.0)	3.3 (0.5)

	Work Together in Small Groups	Solve With Everyday Life Things	Teacher Gives Us Homework	Can Begin Our Homework in Class	Teacher Checks Our Homework	Check Each Other's Homework	Discuss Completed Homework
USA	41.7 (2.1)	52.1 (1.5)	89.1 (1.5)	74.8 (1.4)	75.9 (2.1)	43.3 (2.5)	78.0 (1.6)
Australia	25.5 (1.4)	46.4 (1.0)	85.9 (1.4)	48.2 (1.3)	48.2 (2.0)	15.4 (0.8)	47.4 (1.7)
Belgium (Fl)	6.5 (1.0)	25.5 (1.2)	87.5 (1.9)	32.3 (2.8)	85.3 (1.4)	19.9 (1.2)	55.0 (1.7)
Canada	30.3 (2.1)	50.9 (1.2)	83.7 (2.3)	76.5 (1.7)	62.4 (1.8)	28.6 (1.6)	58.4 (1.7)
Czech Republic	9.9 (0.9)	43.0 (1.2)	63.6 (3.0)	14.4 (1.5)	82.6 (2.3)	23.0 (1.5)	42.3 (2.2)
England	32.2 (1.6)	52.7 (1.4)	93.6 (0.7)	29.3 (1.6)	85.2 (1.2)	20.8 (1.2)	60.6 (1.6)
France	9.2 (1.4)	38.2 (1.7)	89.4 (2.0)	30.3 (1.8)	59.7 (1.9)	41.7 (1.1)	44.4 (1.5)
Germany	18.1 (1.7)	28.5 (1.5)	78.5 (1.9)	30.0 (1.8)	67.4 (1.7)	28.8 (1.2)	67.6 (1.3)
Hong Kong	9.4 (1.3)	28.6 (1.1)	89.3 (1.8)	49.2 (2.5)	72.6 (1.8)	28.6 (2.0)	47.2 (1.8)
Hungary	8.4 (0.8)	23.1 (0.9)	95.6 (0.5)	16.8 (1.1)	68.4 (1.8)	21.7 (1.1)	71.3 (1.7)
Israel	34.1 (3.6)	39.6 (2.2)	91.3 (1.8)	58.4 (3.1)	75.2 (2.9)	9.3 (1.1)	61.2 (1.8)
Japan	14.4 (1.5)	18.0 (0.9)	55.3 (2.8)	24.3 (1.3)	37.1 (2.3)	26.3 (1.0)	21.5 (0.9)
Korea	7.1 (0.7)	18.7 (0.9)	59.6 (2.4)	14.6 (1.0)	47.3 (2.1)	16.6 (1.8)	12.9 (0.8)
Netherlands	14.8 (1.7)	24.5 (1.5)	96.0 (0.9)	81.1 (1.9)	40.1 (2.7)	4.1 (0.5)	79.8 (2.5)
New Zealand	37.9 (2.0)	54.0 (1.1)	84.2 (1.8)	36.8 (1.8)	72.7 (2.0)	13.5 (0.8)	47.9 (1.5)
Norway	20.1 (1.8)	23.8 (1.1)	94.7 (1.0)	39.0 (2.7)	64.9 (2.5)	4.9 (0.5)	34.8 (1.5)
Russian Federation	13.0 (0.9)	30.3 (1.6)	98.5 (0.3)	13.1 (1.0)	73.5 (2.0)	22.4 (1.4)	57.7 (2.4)
Singapore	16.0 (1.5)	39.8 (1.1)	96.8 (0.8)	56.5 (1.8)	91.3 (1.1)	29.8 (0.9)	57.1 (1.7)
Spain	17.0 (1.6)	53.4 (1.6)	94.2 (0.7)	64.7 (1.9)	62.5 (2.1)	13.4 (0.9)	69.4 (1.5)
Sweden	23.0 (1.3)	30.4 (1.1)	72.0 (2.3)	23.8 (1.5)	74.7 (2.4)	6.8 (0.5)	40.0 (2.0)
Switzerland	23.6 (1.5)	32.8 (1.3)	86.8 (1.7)	49.9 (1.7)	63.7 (1.9)	35.9 (1.6)	61.9 (1.4)
Thailand	50.6 (1.1)	37.5 (1.1)	96.6 (0.5)	83.4 (1.0)	84.5 (1.4)	48.6 (1.6)	61.8 (1.4)

Most US seventh and eighth grade mathematics students reported having their homework reviewed. About 75 percent reported that their teacher always or often checked their homework, and about 45 percent reported that other students often or always checked their homework. While in Japan and the Netherlands nearly 40 percent of students indicated teachers always or often checked their homework and less than 30 percent reported often or always checking each others' homework. Further, 78 percent of US mathematics students reported that they always or often discussed their completed homework. Homework was clearly a central object of attention in US seventh and eighth grade mathematics classrooms, much more so than in several other countries.

We found a similar, although less pronounced, preoccupation with homework in seventh and eighth grade science as perceived by US students. Roughly 60 percent report often or always beginning to work on their homework during class, and discussing it in class (see Exhibit 3.7). Given science teacher reports of less use of in-class exercises and more emphasis on new instruction than was the case for mathematics, this homework likely differed in character from the mathematics homework.

Demonstrations and experiments were central features in US eighth grade science classrooms. About 60 to 70 percent of US science students reported always or often having a demonstration or doing an experiment in class. In several countries, including Sweden, New Zealand, Singapore, England, and Hong Kong, 80 percent or more of the students reported doing experiments in class.

Other Instructional Practices. Since the student responses were to somewhat more detailed activities than the teacher responses previously discussed, it is worthwhile to briefly report some additional specifics that teachers reported about classroom practices. Over one-half of Japanese, Korean, Czech and Hungarian eighth grade mathematics teachers reported having students practice writing equations in every lesson or most lessons. Only about one-third of US teachers reported the same. By contrast, nearly 60 percent of the US teachers reported having students practice computations in most lessons or every lesson. In Germany, Korea, and Norway about one-third of the teachers reported the same. In many of the other countries as large (if not larger) a percentage of teachers reported practicing computations regularly. The US classes were clearly computationally-centered as opposed to the algebraically-centered classes of the Czech Republic, Hungary, and Japan – as would be expected given the differences in mathematics curricula for these countries at this level.

In eighth grade science (see Exhibit 3.9) 60 percent or more of the teachers in England, Israel, New Zealand, and Japan reported having students write out their explanations of what they observed in class. In the US, only about one-third of the teachers reported doing so. Science classrooms in these other countries seemed oriented to description and observation. Half or more of the teachers in the Russian Federation, Thailand, and Hungary had students organize or

Exhibit 3.7. *Eighth Grade Students' Perceptions of Classroom Activities in Science (percent responding 'always' or 'often').*

	Teacher Shows Us How to Do Problems	Copy Notes From the Board	Have a Quiz or Test	Work on Our Own	Use Computers	Use Calculators	Work in Small Groups	Solve Problems with Everyday Life Things
USA	69.2 (1.2)	73.4 (1.6)	76.9 (1.4)	79.3 (1.5)	15.1 (1.2)	18.2 (1.3)	64.9 (1.5)	51.0 (0.9)
Australia	69.9 (1.1)	86.5 (1.0)	43.9 (1.2)	67.4 (1.5)	5.9 (0.6)	9.3 (0.7)	64.4 (1.3)	43.4 (0.8)
Belgium (Fl)*	65.3 (2.0)	68.2 (2.4)	78.0 (2.3)	52.1 (2.5)	6.7 (1.7)	33.4 (3.0)	48.8 (3.1)	51.0 (2.2)
Canada	70.8 (1.5)	75.7 (2.0)	60.5 (1.4)	70.3 (1.7)	10.2 (0.8)	15.1 (1.1)	66.9 (1.4)	52.5 (1.1)
Czech Republic**	98.6 (0.3)	83.3 (1.7)	36.6 (2.1)	43.2 (2.5)	0.8 (0.2)	12.2 (1.3)	17.6 (1.3)	31.4 (1.5)
England	86.1 (1.0)	89.7 (1.3)	54.2 (2.0)	47.9 (1.6)	5.4 (0.7)	13.6 (1.4)	84.7 (1.1)	50.7 (1.2)
France*	79.0 (1.6)	77.5 (2.1)	83.3 (1.4)	39.3 (1.4)	6.0 (1.1)	19.0 (1.3)	54.5 (2.4)	51.4 (1.5)
Germany**	62.8 (1.7)	75.5 (1.8)	55.6 (2.2)	33.2 (2.0)	7.3 (1.0)	14.4 (1.6)	29.8 (2.5)	34.5 (1.7)
Hong Kong	80.0 (1.5)	60.9 (2.2)	61.9 (2.6)	55.4 (1.8)	4.6 (0.6)	7.3 (0.8)	58.6 (1.6)	57.3 (1.5)
Hungary**	82.4 (1.2)	83.0 (1.2)	24.8 (1.3)	63.9 (2.0)	3.5 (0.5)	22.5 (2.1)	12.8 (1.0)	29.0 (1.2)
Israel	54.4 (2.2)	78.4 (2.2)	47.0 (2.9)	60.5 (2.4)	11.8 (3.1)	8.2 (0.9)	45.7 (2.6)	39.8 (2.0)
Japan	68.8 (1.0)	92.1 (1.0)	32.4 (2.2)	28.3 (1.3)	2.9 (0.7)	1.6 (0.2)	35.8 (1.7)	22.8 (0.9)
Korea	63.8 (1.4)	87.8 (1.6)	21.9 (1.3)	26.1 (1.0)	3.1 (0.3)	3.1 (0.3)	13.3 (0.8)	16.5 (0.8)
New Zealand	76.9 (1.2)	86.4 (1.2)	49.2 (1.7)	68.3 (1.5)	5.4 (0.8)	8.4 (0.8)	74.2 (1.4)	47.8 (1.1)
Norway	68.6 (1.1)	81.7 (1.4)	45.1 (1.7)	49.1 (1.9)	4.1 (0.6)	6.5 (0.6)	61.0 (1.6)	31.4 (1.0)
Russian Federation**	84.8 (0.9)	84.5 (1.3)	73.3 (1.4)	57.6 (1.5)	5.3 (0.5)	35.8 (1.7)	20.9 (2.0)	32.2 (2.0)
Singapore	78.4 (1.2)	74.7 (2.1)	73.8 (1.4)	70.8 (1.4)	2.0 (0.2)	8.8 (1.1)	50.5 (2.2)	58.8 (1.1)
Spain	86.2 (1.2)	70.7 (1.6)	74.9 (1.4)	67.0 (1.3)	4.3 (0.6)	22.3 (2.4)	30.8 (1.9)	43.9 (1.3)
Sweden**	84.2 (1.0)	77.5 (1.9)	66.8 (1.6)	65.4 (1.8)	6.7 (0.8)	8.8 (0.9)	74.6 (2.0)	43.1 (1.7)
Switzerland	53.8 (1.4)	59.0 (1.9)	49.2 (1.4)	43.2 (1.4)	3.2 (0.4)	5.3 (0.8)	44.5 (2.1)	40.1 (1.1)
Thailand	70.9 (1.4)	83.0 (1.1)	61.7 (1.5)	56.9 (1.1)	4.9 (0.8)	6.0 (0.7)	74.9 (1.1)	47.8 (1.3)

	Can Begin Our Homework in Class	Teacher Checks Our Homework	Check Each Other's Homework	Discuss Completed Homework	Work on Projects	Teacher Gives Us Homework	Teacher Gives a Demonstration	Do an Experiment in Class
USA	59.3 (1.7)	73.2 (1.6)	41.4 (2.2)	63.4 (1.7)	61.3 (1.1)	66.9 (2.2)	68.4 (1.4)	61.6 (1.7)
Australia	33.8 (1.3)	52.5 (1.6)	19.2 (0.8)	43.7 (1.3)	42.8 (1.1)	59.7 (1.7)	75.4 (1.1)	77.0 (1.4)
Belgium (Fl)*	20.8 (2.9)	53.3 (2.6)	18.5 (1.7)	38.4 (2.8)	60.5 (2.3)	42.3 (4.1)	84.7 (2.7)	71.6 (2.9)
Canada	56.9 (1.5)	63.5 (1.9)	28.5 (1.3)	55.7 (1.3)	62.2 (1.4)	56.1 (1.8)	72.6 (1.5)	69.7 (1.8)
Czech Republic**	15.2 (1.2)	44.7 (2.3)	20.4 (1.1)	33.4 (1.6)	19.0 (1.3)	18.1 (1.7)	70.1 (2.5)	34.8 (2.2)
England	31.6 (1.5)	90.3 (1.0)	8.3 (0.8)	49.8 (1.7)	55.0 (2.2)	90.4 (0.9)	90.2 (0.9)	91.2 (0.6)
France*	25.3 (1.8)	64.6 (2.5)	39.0 (1.5)	48.6 (1.9)	54.8 (1.4)	71.6 (2.9)	89.8 (1.1)	73.7 (2.0)
Germany**	21.3 (1.5)	52.8 (3.0)	24.0 (1.5)	40.9 (2.3)	48.5 (2.1)	41.5 (3.0)	75.8 (1.8)	47.7 (3.1)
Hong Kong	31.0 (1.7)	57.4 (2.2)	26.7 (1.4)	38.2 (1.5)	28.2 (1.9)	58.1 (2.5)	91.0 (1.1)	83.3 (2.0)
Hungary**	19.8 (1.2)	52.5 (2.1)	19.6 (1.0)	54.2 (1.8)	58.4 (1.1)	59.8 (2.2)	80.5 (1.7)	19.9 (1.6)
Israel	56.1 (2.9)	80.4 (2.4)	11.3 (1.1)	65.8 (2.4)	30.3 (1.9)	86.3 (2.8)	72.8 (2.7)	53.3 (2.8)
Japan	10.0 (0.8)	27.7 (1.9)	14.8 (0.9)	12.3 (0.7)	27.8 (1.1)	26.3 (2.1)	65.6 (1.6)	76.6 (1.5)
Korea	9.9 (0.7)	37.2 (1.9)	12.5 (0.9)	13.9 (0.8)	45.6 (1.5)	28.6 (1.8)	42.3 (1.7)	32.6 (1.7)
New Zealand	28.5 (1.3)	70.1 (1.6)	18.9 (1.1)	47.6 (1.5)	48.3 (1.9)	69.2 (2.0)	78.5 (1.2)	80.8 (1.3)
Norway	33.1 (1.8)	61.0 (2.0)	7.7 (0.6)	39.0 (1.3)	52.4 (1.4)	81.4 (1.5)	70.9 (1.6)	66.3 (2.2)
Russian Federation**	23.0 (1.1)	65.0 (1.2)	28.5 (1.1)	44.4 (2.2)	21.8 (1.1)	82.3 (1.1)	70.6 (1.9)	45.4 (2.4)
Singapore	41.2 (1.6)	79.5 (1.3)	34.9 (1.4)	61.9 (1.5)	21.2 (1.2)	71.2 (2.2)	86.5 (1.0)	84.6 (1.0)
Spain	66.0 (1.9)	63.4 (2.0)	15.8 (0.9)	64.8 (1.5)	45.6 (1.7)	84.5 (1.6)	28.0 (1.8)	23.3 (1.6)
Sweden**	29.7 (1.7)	78.1 (1.8)	20.0 (1.2)	48.2 (1.7)	43.4 (1.4)	75.8 (2.5)	90.4 (0.9)	91.7 (0.8)
Switzerland	27.4 (1.3)	52.3 (1.8)	24.7 (1.1)	44.5 (1.5)	36.9 (1.2)	46.8 (1.9)	50.8 (2.1)	35.5 (1.7)
Thailand	83.8 (1.0)	72.6 (2.0)	48.7 (1.6)	60.2 (1.3)	41.8 (1.8)	75.7 (1.8)	84.1 (1.3)	54.6 (1.2)

* Responses for physics students

** Responses for chemistry students

Exhibit 3.8. *Percent of Eighth Grade Mathematics Teachers Who Ask*
Students to Write Explanations or Organize and Provide
Reasons in 'Most' or 'Every Lesson.'

	Write Equations		Practice Computations	
USA	36.4	(3.9)	59.1	(3.8)
Australia	15.7	(4.6)	50.5	(3.4)
Belgium (Fl)	6.4	(3.5)	67.1	(4.0)
Canada	28.2	(0.9)	60.1	(4.2)
Czech Republic	61.8	(4.3)	77.0	(4.8)
England	21.1	(3.8)	41.6	(2.8)
France	8.3	(1.8)	50.8	(4.6)
Germany	30.3	(2.5)	31.4	(5.1)
Hong Kong	40.0	(4.1)	55.3	(5.6)
Hungary	67.7	(3.8)	86.7	(3.1)
Israel	44.0	(7.5)	45.3	(7.0)
Japan	71.7	(4.1)		
Korea	55.2	(5.9)	28.0	(4.2)
New Zealand	9.9	(4.8)	47.1	(3.8)
Norway	3.5	(2.9)	35.9	(4.5)
Russian Federation	31.4	(2.9)	86.8	(2.5)
Singapore	24.8	(5.9)	49.7	(4.7)
Sweden	1.6	(3.8)	79.9	(2.6)
Switzerland	16.2	(2.0)	75.2	(4.1)
Thailand	17.2	(2.8)	86.8	(4.6)

Exhibit 3.9. *Percent of Eighth Grade Science Teachers Who Ask Students to*
Write Explanations or Organize and Provide Reasons in 'Most'
Or 'Every Lesson.'

	Write Explanation		Organize & Provide Reasons	
USA	36.1	(4.4)	27.2	(3.9)
Australia	49.2	(3.9)	9.1	(1.9)
Belgium (Fl)	16.5	(3.8)	12.3	(2.9)
Canada	53.9	(3.9)	17.4	(3.3)
Czech Republic	46.9	(3.7)	27.1	(2.9)
England	72.2	(3.0)	22.0	(2.9)
France	31.7	(3.2)	25.2	(3.3)
Germany	38.2	(4.8)	37.9	(4.8)
Hong Kong	57.2	(5.1)	28.4	(5.3)
Hungary	38.3	(2.3)	50.1	(2.3)
Israel	65.9	(7.0)	38.5	(10.3)
Japan	59.5	(4.1)	39.9	(3.9)
Korea	49.4	(4.4)	34.4	(3.6)
Netherlands	28.3	(2.7)	12.3	(2.5)
New Zealand	69.0	(3.7)	11.2	(2.9)
Norway	22.9	(4.7)	1.4	(1.4)
Russian Federation	16.7	(1.9)	60.4	(4.1)
Singapore	50.0	(4.8)	24.8	(4.1)
Spain	26.1	(4.3)	20.0	(4.0)
Sweden	46.5	(3.9)	11.6	(2.3)
Switzerland	29.5	(4.8)	19.1	(4.2)
Thailand	49.6	(6.6)	51.8	(6.6)

provide reasons behind their observations. In the US less than 30 percent of the teachers had students do this.

Assessment Practices. Teachers were also surveyed concerning how much 'weight' they assigned to various means of assessing student work. Exhibit 3.10 reports responses for eighth grade mathematics teachers. Obviously, teachers in each country used a mix of assessment tools or means. As in many other countries, US eighth grade mathematics teachers relied on tests and homework, but less than teachers in many other countries on their observations of students and students' responses in class.

Exhibit 3.10. *Percent of Eighth Grade Mathematics Teachers Emphasizing Various Means in Assessing Student Work.*

	Standardized Tests	Teacher-Made Constructed Response	Teacher-Made Objective	Homework	Projects	Observations	Student Responses in Class
USA	21.4 (2.7)	51.6 (6.2)	26.3 (5.8)	57.0 (7.0)	34.5 (4.7)	45.1 (5.9)	45.7 (6.0)
Australia	7.6 (2.6)	42.5 (5.5)	23.9 (4.1)	26.1 (3.9)	29.3 (3.8)	37.4 (5.1)	33.9 (4.8)
Belgium (Fl)	10.0 (3.4)	94.3 (9.7)	11.3 (4.0)	15.0 (3.9)	15.6 (2.7)	50.0 (6.6)	54.8 (6.7)
Canada	16.3 (4.4)	48.8 (7.4)	18.3 (3.3)	44.3 (6.3)	31.7 (5.4)	43.3 (6.7)	41.4 (6.2)
Czech Republic	42.6 (9.4)	99.6 (8.0)	19.4 (6.4)	13.6 (3.1)	29.4 (5.7)	74.3 (8.3)	96.2 (10.2)
England	35.8 (4.8)	32.5 (4.7)	6.9 (2.4)	68.5 (5.4)	48.4 (4.7)	70.9 (5.7)	66.5 (5.7)
France	22.7 (5.8)	83.5 (9.0)	24.9 (5.7)	28.5 (7.3)	15.5 (5.1)	49.4 (8.4)	54.3 (9.2)
Germany	0.0	54.7 (8.4)	7.2 (3.7)	18.2 (5.5)	40.1 (6.8)	73.7 (10.1)	80.9 (10.2)
Hong Kong	32.0 (8.0)	39.8 (7.9)	40.3 (9.4)	74.1 (10.5)	12.2 (3.7)	68.4 (9.0)	73.8 (9.5)
Hungary	34.3 (5.7)	71.1 (7.2)	24.4 (4.4)	42.6 (6.8)	89.7 (8.8)	68.8 (7.7)	86.7 (8.3)
Israel	77.5 (14.4)	28.9 (7.4)	63.8 (12.7)	60.6 (12.0)	70.2 (14.7)	54.0 (12.5)	46.2 (9.1)
Japan	15.9 (4.2)	54.2 (7.8)	19.9 (4.9)	44.5 (5.8)	34.2 (5.7)	68.0 (8.2)	71.1 (7.7)
Korea	36.2 (5.6)	53.6 (6.7)	32.2 (4.8)	23.9 (4.6)	20.4 (4.5)	30.8 (5.2)	62.4 (6.9)
Netherlands	29.2 (8.0)	98.9 (12.4)	31.4 (9.0)	30.1 (6.6)	14.5 (4.1)	36.4 (7.4)	41.7 (7.7)
New Zealand	13.7 (4.2)	52.5 (6.2)	20.1 (4.5)	33.7 (4.8)	36.4 (5.7)	52.4 (6.3)	46.4 (5.7)
Norway	27.1 (5.9)	100.0		2.7 (1.6)	24.8 (5.6)	14.9 (4.6)	54.9 (6.9)	59.1 (7.6)
Russian Federation	0.0	99.6 (7.9)	53.9 (7.7)	63.6 (8.4)	51.7 (9.4)	96.6 (10.1)	0.0
Singapore	0.0	29.9 (5.6)	6.3 (2.7)	71.7 (8.9)	37.0 (7.2)	60.7 (8.6)	70.1 (8.4)
Spain	4.6 (3.0)	91.5 (8.5)	22.7 (5.1)	74.6 (7.7)	41.6 (8.3)	89.8 (8.5)	95.4 (7.8)
Sweden	59.3 (5.5)	89.9 (6.7)	18.7 (4.1)	50.2 (5.3)	52.5 (4.8)	87.4 (6.4)	78.7 (5.0)
Switzerland	28.2 (5.6)	76.5 (8.8)	6.4 (2.5)	13.2 (2.9)	14.2 (3.5)	47.3 (6.7)	54.3 (6.1)
Thailand	21.8 (7.6)	52.2 (9.4)	71.3 (10.7)	74.7 (10.0)	20.7 (6.1)	51.2 (11.0)	65.7 (10.1)

In eighth grade science classrooms, the pattern was similar as Exhibit 3.11 on page 77 shows. For US teachers, tests and projects dominated assessment. In the Czech Republic, tests and student class responses dominated. Japanese teachers reported relying more on projects, observations of students, and on tests. In all of the countries, science teachers at this level tested more than did their counterparts in mathematics.

Exhibit 3.11. *Percent of Eighth Grade Science Teachers Emphasizing Various Means in Assessing Student Work.*

	Standardized Tests	Teacher-Made Constructed Response	Teacher-Made Objective	Homework	Projects	Observations	Student Responses in Class
USA	13.1 (3.6)	64.6 (7.9)	72.5 (8.1)	53.4 (5.8)	77.2 (5.3)	46.2 (6.8)	36.8 (6.7)
Australia	5.5 (2.0)	70.8 (6.2)	60.4 (5.7)	45.6 (5.6)	67.8 (5.4)	41.5 (4.9)	25.6 (4.1)
Belgium (Fl)	11.1 (5.3)	91.9 (9.1)	28.1 (6.0)	20.0 (6.7)	39.1 (6.1)	47.7 (6.8)	50.3 (6.4)
Canada	8.1 (2.0)	75.0 (8.1)	49.1 (6.2)	50.1 (6.5)	76.0 (7.2)	36.3 (5.7)	31.5 (6.2)
Czech Republic	40.5 (4.1)	93.4 (4.7)	36.7 (4.2)	9.9 (1.9)	48.2 (5.0)	72.0 (5.7)	94.1 (6.2)
England	39.0 (4.7)	68.1 (5.3)	19.6 (3.0)	66.2 (3.6)	74.0 (5.2)	64.8 (5.6)	61.0 (4.8)
France	20.3 (3.6)	88.8 (6.7)	43.8 (6.0)	37.4 (5.1)	50.7 (5.6)	71.1 (6.7)	67.4 (7.0)
Germany	5.4 (2.5)	84.3 (7.7)	9.9 (3.0)	30.1 (4.4)	55.1 (6.1)	72.4 (7.8)	86.3 (8.8)
Hong Kong	21.7 (6.6)	49.0 (8.7)	78.1 (9.5)	53.2 (7.8)	41.1 (6.7)	42.9 (7.4)	42.9 (6.9)
Hungary	45.7 (4.3)	89.5 (5.3)	35.6 (3.5)	42.3 (3.9)	81.5 (4.9)	71.2 (4.2)	88.5 (5.3)
Israel	20.7 (10.7)	68.9 (14.1)	92.2 (18.2)	34.7 (10.1)	47.6 (12.7)	59.7 (16.1)	70.6 (17.4)
Japan	16.2 (4.3)	71.9 (7.5)	44.8 (7.5)	44.5 (6.7)	87.7 (9.0)	78.8 (7.9)	68.9 (8.2)
Korea	22.7 (5.7)	41.1 (5.2)	40.9 (6.1)	15.9 (4.8)	54.8 (6.0)	38.1 (6.1)	38.1 (6.1)
Netherlands	60.2 (6.8)	90.2 (6.9)	63.9 (6.3)	11.5 (3.7)	25.4 (5.0)	16.6 (3.3)	13.8 (3.4)
New Zealand	9.7 (3.5)	63.1 (5.4)	55.9 (6.0)	30.4 (4.7)	65.8 (6.0)	52.7 (6.6)	35.7 (5.7)
Norway	5.6 (2.5)	95.0 (10.1)	7.7 (2.9)	56.5 (7.8)	68.3 (9.3)	67.8 (8.6)	74.1 (8.9)
Russian Federation	0.0	95.6 (5.2)	63.3 (4.9)	77.0 (5.6)	73.9 (6.2)	96.7 (6.5)	0.0
Singapore	0.0	80.0 (8.6)	61.3 (7.5)	47.5 (6.4)	77.0 (7.7)	47.4 (7.1)	46.3 (7.5)
Spain	7.9 (3.3)	96.5 (7.3)	43.2 (6.6)	75.5 (7.6)	62.2 (8.6)	88.0 (8.6)	92.1 (8.7)
Sweden	13.1 (4.0)	87.2 (6.5)	21.9 (3.9)	70.4 (5.6)	95.1 (6.1)	88.1 (6.6)	92.2 (6.3)
Switzerland	10.6 (3.4)	88.1 (8.8)	20.4 (5.2)	13.4 (4.4)	46.4 (7.0)	53.7 (7.7)	61.4 (8.2)
Thailand	20.4 (7.5)	63.1 (10.0)	81.4 (11.5)	64.2 (9.8)	70.1 (10.1)	66.6 (10.6)	67.8 (11.2)

OTHER ASPECTS OF INSTRUCTIONAL PRACTICE: A SUMMARY

Among other things that were noted about instructional practices were the following:

- What US students said about their mathematics instruction was consistent with teacher descriptions of their instruction. Students said they often worked on their own in class and that they often began homework in class. They also indicated that considerable class time was spent reviewing their homework. US mathematics classrooms centered around homework far more than did the classrooms of several other countries.

- What US students said about their science instruction was fairly similar to what they said about their mathematics instruction, although teacher reports of classroom practices differed somewhat for the two areas. The differences teachers reported were apparently sufficiently slight or subtle that student questionnaire responses did not reflect them as clearly as the teacher questionnaire responses did.

Further characteristics of US eighth grade mathematics instruction included:

- Giving more tests than did Japanese, Korean, Czech and German mathematics teachers.

- Often having students work in groups (about 40 percent of US students reported this compared to 20 percent or less of students in most other countries including Japan, Korea, the Czech Republic, Hong Kong, and Singapore).

- Emphasizing computation heavily.

- Far less emphasis on writing equations.

- Using tests and homework primarily in assessing student progress rather than other tools such as observing students or student projects.

US eighth grade science instructional practices can be characterized as:

- Having often involved students in doing experiments in class or having a class demonstration by the teacher (as opposed to other classrooms such as those in Hong Kong, Japan, and Singapore which were even more experiment-centered).

- Not often asking students to write out explanations (in marked contrast to English, Japanese, Israeli, and New Zealander common practice).

- Providing more direct instruction on new materials than was true for mathematics.

- Using projects as well as tests to assess student progress.

CONCLUDING REMARKS: US SCIENCE AND MATHEMATICS INSTRUCTION

Four approaches were taken to characterize instructional practices (three of which served as the basis for data reported in this chapter and one for data reported in the next chapter). Teachers were asked through survey questionnaires to describe specific instructional practices and behaviors. A category system (typology) was formed for classifying how teachers structured their lessons based on their own reports of how they used different activities. A category system was formed to characterize a teacher's pedagogical approach based on their reported beliefs about teaching practices and about the subject matter that they taught. Students were asked to report how they perceived the instruction they received (which provided both 'triangulation' on the accuracy of teacher self-reports and additional insight into differing instructional practices). These four approaches were used to describe and analyze instructional practices in US mathematics and science instruction, and to compare those practices to their counterparts in other countries.

US mathematics classrooms seem dominated by seatwork and homework. Since this differed from the case for science in the US, this seems more likely to have been a characteristic of how school mathematics was conceived as opposed to school science, rather than a characteristic national pattern. Comparatively less time was devoted to instruction on new material and more was devoted to review and discussion of homework. Instruction was present in US lessons but not as dominantly as in Japanese lessons.[6] Homework was frequently started during class. Seatwork was not typically an integral part of class discussion in the US (in contrast to Japan). Student responses confirmed this emphasis on homework and its dominance in class.

According to students, US eighth grade mathematics teachers also gave more tests than teachers in other countries. They made somewhat more use of group work and relied more on tests and homework in assessing students. US classrooms emphasized computation rather than writing equations (as was the case in Japan).

Science instruction differed from mathematics instruction and not just in its emphasis on homework. Science instruction centered on instruction of new subject matter content perhaps in combination with seatwork.

US eighth grade science classrooms made comparatively more use of demonstrations, although there was some use of experiments as well (comparing to the more exclusive emphasis on experiments characteristic of Japan). They focused less on having students write out explanations of what they observed. They tended to supplement tests with projects rather than homework in assessing students.

It is often hard to find the 'systematic' in such a plethora of details. However, it appears that in US mathematics classrooms the subject matter was treated consistently as if it had a different structure than appeared to be the case in some of the high achieving countries given their instructional patterns. US mathematics instruction was centered on computation and homework. It focused comparatively less on new instruction and more on review. It used more 'discretized' assessment tools (tests and homework) rather than more integrated, qualitative methods (observing students, evaluating student projects). These practices are consistent with a view of school mathematics as existing in discrete 'bits', mastered cumulatively and incrementally, for which constant review is necessary and for which constant 'objective' monitoring is needed to assess the state of assimilation and mastery.

This seems consistent with the type of content taught as described in the previous chapter. It is important not to isolate factors since some will note that some of the top achieving countries in mathematics had considerable review and seatwork. However, putting these factors together with what they taught as seen in the previous chapter makes it clear that their approach was not closely similar to the US'. In contrast, their approach took place with fewer topics and a typically more demanding curriculum at seventh and eighth grade. These were national level, systemic characteristics.

There appears to have been no counterpart to the 'discretization' and incremental mastery assumptions in US science instruction. This allowed relatively more emphasis on instruction on new material and a much smaller role for homework. Assessment could focus more on student projects as well as tests. US science instruction, however, appears to have been more dominated by a mixture of classroom demonstrations and experiments rather than just experiments, and not to have focused on writing explanations for what was observed in classroom activities. These are characteristics of a more 'reception learning' approach only somewhat modified by active experimentation and project work.

Clearly there were dominant patterns characteristic of entire groups of US science and mathematics teachers, so we cannot regard instructional practices solely as the result of individual teacher choices. They reflected systemic differences, apparently in how school mathematics and school science were conceived. The operative conceptions seem to have remained more traditional, 'reception learner' views. Those who desire US science and mathematics classrooms to go 'back to the basics' are making a rash assumption unsupported by data that our classrooms have ever strayed far from their traditional roots.

Notes–

[1] See Schmidt, W.H. et al. (1966). *Characterizing Pedagogical Flow: An Investigation of Mathematics and Science Teaching in Six Countries.* Dordrecht: Kluwer.

[2] See Schmidt, W.H. et al. (1966). *Characterizing Pedagogical Flow: An Investigation of Mathematics and Science Teaching in Six Countries.* Dordrecht: Kluwer.

[3] Teachers at this grade level were not classified into mathematics or science teachers since in the US (as well as in most TIMSS countries) most teachers taught both subjects.

[4] The former two patterns were more characteristic of the US and the latter of Japan. See Schmidt, W.H. et al. (1966). *Characterizing Pedagogical Flow: An Investigation of Mathematics and Science Teaching in Six Countries.* Dordrecht: Kluwer .

[5] See Stigler, et al. (in preparation). *The TIMSS Classroom Videotape Study.* Department of Education. National Center for Education Statistics: Washington, DC.

[6] This seems confirmed by the work done by Stigler and others. See Stigler, et al. (in preparation). *The TIMSS Classroom Videotape Study.* Department of Education. National Center for Education Statistics: Washington, DC. See also: http://nces.ed.gov/timss/video/index.html.

Chapter 4
Schools, Teachers, Students, and Other Factors

The last two chapters discussed what we teach in mathematics and the sciences, how we teach it, and to whom we teach it. These factors of how we implement mathematics and science curricula seem clearly to be largely systemic, and not simply the choices of individual teachers. It is possible that they are related to common characteristics of US educational systems. Whether this is so or not, such factors are not the only systemic factors reflected in the data regarding schools, teachers, and students. This chapter examines some of these other factors.

OTHER FEATURES OF TEACHERS AND SCHOOLS

We consider two broad categories of factors: other features of US schools and their science and mathematics teachers, and factors in US schools and students that complicate access to learning possibilities. First, there are many aspects of our formal education systems that likely affect how well we accomplish the goals of schooling. These seemingly systemic factors include school staffing patterns and other school characteristics such as the functions of school principals and even how schools are conceived as institutions. They also include policies on teacher collaboration and teachers' views and beliefs about subject matter. All of these can have an impact on access to learning possibilities in mathematics and the sciences and possibly on resulting aggregate achievements.

TEACHERS' BELIEFS, APPROACHES, AND SPECIALIZATION

What teachers teach and how they teach it are affected by their subject matter beliefs and preferred pedagogical approaches, things that are consequences of their training and experiences. These systemic factors also include the extent to which science and mathematics teachers are allowed to function as either official or *de facto* specialists, or whether they typically have more general assignments. These two sets of factors are discussed here.

Teacher beliefs and pedagogical approaches. What and how teachers teach are affected not only by their own beliefs – by their conceptions of subject matter disciplines in science and mathematics, but also by their beliefs about their students, and by their understanding of appropriate pedagogy. They are also likely the consequences of more systemic educational choices affecting who becomes a teacher, how they are trained, and what their experiences are. More obviously, whether mathematics and science are taught by those who are in some sense 'specialists' in that field is also a characteristic of US education systems.

We mentioned in the previous chapter that we also constructed a typology of teacher beliefs that was based on beliefs about their subject matter discipline (mathematics or science) and about their discipline as represented in school (school science or mathematics). This section discusses that categorization of teachers.

Based on teachers' responses to these items surveying their beliefs, four categories were constructed empirically into which teachers could be classified. The first group of teachers might be characterized as *discipline-oriented teachers*, understanding 'discipline' in the sense of subject matter. Teachers in this group took a more formal, as opposed to a real-world, view of their disciplines.

In mathematics, these teachers more often indicated that it was important to remember formulas, that mathematics was essentially abstract, and that mastering algorithms and basic computation was important. They more often indicated that the 'real world use' of mathematics was less important. They also tended to indicate that success in mathematics learning was more a matter of natural talent than other factors. Science teachers in this group displayed general characteristics similar to those of mathematics teachers, but also more often indicated that creativity was not particularly important in learning science.

The second group might be considered *process-oriented teachers*. In general, they did not take a formal, abstract view of their discipline, but indicated that the real-world use of their discipline was important. They also tended to emphasize creativity and to think about mathematics more conceptually.

Mathematics teachers in this group indicated that it was relatively less important to remember formulas, to focus on algorithms, or to emphasize basic computations. They held that mathematics was not abstract, and that its real-world use was more important. Unlike the discipline-oriented teachers, they indicated that doing well in mathematics was not particularly a matter of natural talent. Science teachers in this group, in addition to sharing characteristics similar to the mathematics teachers, indicated explicitly that creativity was important in mastering science.

The third group shared most things in common with the first group, but were more concerned with emphasizing the real-world use of mathematics.

This group is perhaps best characterized as *procedure-oriented teachers*, concerned with mathematics as a discipline, but one that employs useful representations of the real world. These teachers did not hold as consistently a formal view of their discipline as was the case for discipline-oriented teachers. For example, mathematics teachers in this group regarded algorithms as only modestly important and indicated that subject matter should be presented more conceptually. Mathematics teachers in this group indicated that mastering mathematics was mostly a matter of natural talent. Science teachers tended to indicate that creativity was moderately important.

These emphases, along with their generally more moderate level of choices, suggest that this group viewed their disciplines less holistically than the discipline-oriented teachers. Rather than seeing the discipline as an entire 'gestalt', they may well have seen the components of school mathematics or school science (computation, algorithms, etc.) and reacted strongly to some but not to others. For this reason, we have characterized these teachers as procedure-oriented rather than discipline-oriented since they appear to have focused more on component procedures aimed at certain real world uses (such as seeing computation as more important than the formal structure of algorithms) rather than at the discipline as a whole. We use 'procedures' in comparison to 'process' to indicate their comparatively more formal, reception-learning, and rote approach compared to those in the process group who emphasized ways of thinking.

The fourth group is quite different. They tended to indicate a high level of importance for most things. They felt that it was important to remember formulas, focus on algorithms, and master basic computations. They preferred a more formal view of their discipline. Mathematics teachers in this group indicated that mathematics should be primarily presented as an abstract discipline and they emphasized natural talent. Science teachers in this group did not view creativity as important.

Perhaps the most appropriate label for this group is *eclectic teachers*. By appearing to emphasize everything, they essentially emphasized nothing and did not posses a distinctive character. We use 'eclectic' in this sense of blending, without distinction, elements of all approaches. They were both somewhat discipline-oriented and somewhat real-world-oriented, but also seemed to favor more conceptual process-oriented approaches in some cases. We must also keep in mind that this may not represent a real characteristic pattern of teaching behavior (since it lacks the internal consistency especially of the first three categories) but may be an artifact of these teachers' responses to the survey instruments. However, it was a categorization that held across many countries and cultures some of whom (for example, Japan) have been seen not to be prone to overmarking. In view of that, it seems more likely that this represents a genuine distinct type of teacher.

How were sampled mathematics teachers distributed among these groups? Over one-third of those teaching US thirteen-year-olds fell into the eclectic category that emphasized both a formal view as well as real-world applications. Around 40 percent or more of the teachers from Singapore, Spain, and Thailand fell into this eclectic category. Typically, less than one third of the teachers in most TIMSS countries were classified in this way. Exhibit 4.1 presents a more complete display of these data.

Exhibit 4.1. *Percentage of Seventh- and Eighth-Grade* Mathematics Teachers** with a Particular Pedagogical Orientation.*

	Process-Oriented	Discipline-Oriented	Procedure-Oriented	Eclectic
USA	23.1 (4.7)	11.6 (2.4)	23.1 (5.9)	42.2 (6.0)
Australia	29.4 (2.7)	26.9 (2.6)	17.7 (2.3)	26.0 (2.6)
Belgium (Fl)	5.7 (1.8)	63.5 (3.5)	7.6 (1.6)	23.2 (2.9)
Canada	25.9 (2.8)	14.2 (2.2)	23.2 (3.0)	36.8 (3.2)
Czech Republic	32.5 (4.1)	35.1 (4.5)	18.5 (3.1)	13.8 (2.8)
England	22.9 (4.1)	36.2 (4.2)	15.2 (3.3)	25.7 (3.9)
France	28.1 (3.0)	41.1 (3.5)	17.4 (2.6)	13.4 (2.7)
Germany	31.0 (7.3)	26.1 (5.7)	19.3 (7.6)	23.6 (7.7)
Hong Kong	12.5 (2.8)	42.4 (4.8)	12.1 (2.6)	33.0 (4.4)
Hungary	18.1 (2.5)	24.8 (3.1)	23.0 (3.0)	34.1 (3.4)
Israel	5.1 (2.4)	70.9 (10.4)	5.5 (3.0)	18.5 (10.0)
Japan	11.1 (2.0)	42.3 (3.2)	22.4 (2.9)	24.2 (3.1)
Korea	4.6 (1.2)	53.0 (3.6)	11.8 (1.8)	30.6 (3.3)
Netherlands	29.1 (6.8)	37.9 (6.9)	19.2 (5.5)	13.7 (5.0)
New Zealand	33.2 (3.3)	23.2 (2.9)	22.8 (3.0)	20.8 (2.6)
Norway	20.9 (3.6)	22.1 (2.9)	31.0 (3.5)	26.1 (3.5)
Russian Federation	13.5 (2.7)	40.3 (4.5)	21.4 (3.9)	24.9 (3.7)
Singapore	9.4 (1.8)	33.8 (3.1)	14.6 (2.1)	42.2 (3.3)
Spain	31.1 (4.0)	10.2 (2.4)	19.1 (3.3)	39.5 (4.0)
Sweden	48.0 (3.1)	18.5 (2.2)	22.2 (2.6)	11.2 (1.7)
Switzerland	33.1 (5.5)	43.8 (8.0)	2.5 (0.5)	20.5 (7.2)
Thailand	2.4 (1.1)	30.3 (4.2)	7.3 (2.1)	60.0 (4.6)

* See note 3 on page 12.
** See note 4 on page 61.

About 12 percent of US teachers of thirteen-year-olds were classified as discipline-oriented. They had a formal view of mathematics that did not depend on real world applications. By contrast, almost half or more of the teachers in Hong Kong, Japan, Korea, Belgium (Flemish), and Switzerland fell into this category, but less than 15 percent of the teachers from Canada and Spain. Across all TIMSS countries, about one-third or more of the teachers were classified as belonging to this category.

Almost 25 percent of US seventh and eighth grade teachers were classified as process-oriented teachers. In contrast, only about 10 percent or less of those in Belgium (Flemish), Hong Kong, Israel, Japan, Korea, Singapore, and Thailand fell into this category as well as 30 percent or more of the teachers from most other countries.

Exhibit 4.2. *Percentage of Seventh- and Eighth-Grade Science Teachers with a Particular Pedagogical Orientation.*

	Process-Oriented	Discipline-Oriented	Procedure-Oriented	Eclectic
USA	20.9 (3.8)	8.3 (2.1)	25.3 (3.8)	45.8 (3.1)
Australia	16.2 (5.5)	20.2 (5.4)	20.7 (5.8)	43.0 (7.1)
Belgium (Fl)	23.6 (3.8)	40.5 (3.6)	19.2 (3.6)	16.7 (2.5)
Canada	30.4 (5.2)	5.3 (1.5)	21.3 (4.2)	43.1 (5.2)
Czech Republic	30.1 (2.1)	35.7 (2.2)	24.4 (2.4)	9.8 (1.6)
France	44.4 (3.6)	21.2 (2.6)	16.9 (2.6)	17.4 (2.6)
Germany	11.6 (5.4)	38.9 (10.8)	24.0 (9.9)	25.5 (10.5)
Hong Kong	11.8 (2.7)	26.3 (3.5)	26.8 (3.4)	35.1 (4.7)
Hungary	18.5 (2.1)	25.8 (2.5)	24.9 (2.1)	30.9 (2.4)
Israel	18.1 (12.0)	29.1 (6.1)	14.7 (10.9)	38.0 (16.0)
Japan	21.1 (2.8)	35.1 (3.0)	21.2 (2.3)	22.7 (2.9)
Korea	22.4 (3.7)	35.1 (4.1)	22.2 (3.3)	20.3 (3.1)
Netherlands	27.7 (2.6)	31.4 (2.6)	22.2 (2.1)	18.7 (2.4)
New Zealand	33.4 (5.3)	16.9 (4.2)	23.2 (4.4)	26.5 (4.4)
Norway	39.1 (4.2)	21.9 (3.7)	24.9 (3.3)	14.1 (2.9)
Russian Federation	18.0 (2.7)	32.9 (2.5)	27.9 (3.5)	21.3 (2.7)
Singapore	5.0 (1.5)	18.3 (2.5)	23.7 (2.7)	53.1 (2.9)
Spain	29.9 (4.3)	8.4 (2.2)	17.6 (2.9)	44.1 (4.3)
Sweden	58.7 (3.6)	17.7 (2.4)	15.7 (2.1)	7.9 (2.1)
Switzerland	52.8 (8.0)	15.2 (4.3)	17.1 (6.6)	14.9 (7.0)
Thailand	18.1 (3.8)	9.9 (3.6)	13.5 (3.5)	58.5 (5.0)

Not quite 25 percent of US seventh and eighth grade teachers were classified as procedure-oriented teachers. By comparison, a smaller percentage of the teachers from all of the other countries fell into this category with the exception of Norway, where 31 percent were classified as procedure-oriented.

Science teachers for seventh and eighth grade students were similarly distributed among the four categories. As for the US mathematics teachers, nearly one-half fell into the category of eclectic teachers. They viewed science formally, but also emphasized the importance of real-world applications and creativity. Again, about one-half or more of teachers from Spain, Singapore, and Thailand fell into this category.

Less than 10 percent of US thirteen-year-olds' science teachers fell into the discipline-oriented category. Again, in contrast, over one-third of Japanese thirteen-year-olds' science teachers fell into this category, as well as around one-fourth to one-third of the teachers in most other TIMSS countries.

Another 20 percent of US science teachers at this level fell into the process-oriented teacher category. In contrast, nearly twice as many (around 40 percent or more) of the French, Swedish, Swiss and Norwegian science teachers fell into this category, but only slightly more than 10 percent of those from Hong Kong and Germany.

About one fourth of the US science teachers fell into the procedures-oriented category. This compared with an average of around 20 percent across all other TIMSS countries.

Overall for thirteen-year-olds' mathematics teachers, our eclectic teacher category was widely represented in all countries — from as little as 11 percent in Sweden to a high of 60 percent in Thailand. Discipline-oriented teachers were a consistently significant proportion of those classified in each country (from about 20 to a high of almost 65 percent in Belgium (Flemish) and 71 percent in Israel). Procedure-oriented teachers were consistently at modest levels (around 20 percent or less, except in Norway for which they represented about one-third of the responding teachers). Notable exceptions were Spain and the US where only about 10 percent of the teachers fell into this category. Process-oriented teachers were modestly represented in all countries except New Zealand, Switzerland, Sweden, Germany, and the Czech Republic for which they represented one-third or more of those classified.

Overall for thirteen-year-olds' science teachers, eclectic teachers were somewhat more strongly represented than in mathematics (from around 10 to 60 percent among the countries). Process-oriented teachers were the next most dominant (from less than 10 to a high of almost 60 percent in Sweden) except in Singapore. Discipline-oriented teachers were consistently modestly represented (from about 10 to 40 percent, this higher number being true for Belgium (Flemish)). Only in Belgium (Flemish) and the Czech Republic did the procedure-oriented teachers outnumber the eclectic teachers responding (about 19 versus 17 percent and about 25 versus 10 percent, respectively).

What can we say about US seventh and eighth grade mathematics and science teachers as a whole? The very fact that we could clearly distinguish four groups empirically suggests that there were distinct differences among teachers. Furthermore, US mathematics teachers at this level were spread across these categories with the least common category having around 10 percent and the most common over 40 percent of the teachers. By contrast, almost half or more of the Belgian (Flemish), Israeli, Korean, Swedish, and Thai teachers fell into a single category. US science teachers at this level were similarly spread across the categories, with around 45 percent of the teachers falling into the eclectic teacher category. More than half of the science teachers from Sweden, Switzerland, Singapore, and Thailand fell into a single category. In several countries — Belgium (Flemish), Hungary, Japan, Singapore, Spain, and Thailand — the largest proportion of both mathematics and science teachers fell into the same pedagogical orientation category. In fact, of the twenty-one countries represented in the two exhibits for mathematics and science, the dominant pedagogical orientation category was the same in fourteen countries.

Among the implications of this are that US mathematics teachers were divided in their views of mathematics as a discipline, at least as it was to be presented to school children. Nearly three-quarters held a formal view of mathematics as a discipline or at least of how it was to be presented. The comparable number for US science teachers at this level was about 80 percent so they, too, took a predominantly formal approach to science as it was to be presented in school. Factors used to determine how formal the teachers' conceptions of their disciplines were included the belief that the disciplines were primarily abstract and the importance placed on various learning activities such as remembering formulas, learning algorithms, etc. This view of mathematics and science teaching as 'knowledge' to be 'transferred' to students seems to have been a dominant view while a splintered vision of educational goals and contents held sway. This may well have exacerbated the lack of cohesion as varying views of the knowledge to be learned were transmitted to students.

The US was not alone in its teachers' orientation to formal discipline. In Japan, 80 to 90 percent of their counterparts shared similar views. The similarity between the US and Japanese teachers ended, however, at this global level. The largest proportion of both mathematics and science teachers fell into the discipline-oriented category. Nearly one-half of US seventh and eighth grade mathematics and science teachers were in the category of eclectic teachers. Fewer than 25 percent of both science and mathematics teachers in Japan were in the eclectic category that also emphasized a formal view of the discipline.

Most Japanese teachers, falling into the discipline-oriented category, combined their formal views of mathematics and science with less belief in the importance of creativity and attached less importance to real-world uses of the disciplines. By contrast, US mathematics teachers differed further, the largest proportion being classified as eclectic (over 40 percent) and the smallest proportion as discipline oriented (almost 12 percent). Thus, many US mathematics teachers combined their beliefs in presenting mathematics as a formal discipline with beliefs in the importance of emphasizing its real-world uses. In fact, across all categories, teachers with this combination of beliefs amounted to more than half of US mathematics teachers at this level.

Why belabor these distinctions in teacher beliefs? If the teachers' self-reports are at all representative of their actual working beliefs, these are the sort of beliefs that are virtually certain to affect their instructional decisions. If they conceive their discipline as a formal structure or believe it should be presented as such in its school version, they will certainly adopt one set of pedagogical approaches and emphases. If they do not, other approaches and emphases are more likely. If they believe in emphasizing real-world uses, remembering formulas, or emphasizing computation, they are more likely to reflect these things in what they have their students do than those who do not believe or believe less strongly in any of these things. More subtly, if they believe that creativity is less important, or that success is strongly related to natural talent, they are more likely to create similar attitudes in their students or to spend differing amounts of time in the ways they attempt to motivate students.

Put simply, teachers' beliefs about subject matter and appropriate pedagogy have clear empirical consequences so strong that group membership can account for much of the variance in content coverage, emphasis, and many other differences in teacher practices. Further, since teachers, especially in the

US, fall into a relatively small number of categories, this suggests that some aspects of training and experience that cut across state and local lines have determining effects on teachers' practices. These results, if confirmed by further analyses, would open the way to important investigations of how teachers are brought to do the things they do and the effects of teacher education.

In mathematics there appears to be a clear relationship between teacher beliefs and what they do in classrooms. For example, 78 percent of US mathematics teachers in the process-oriented group asked students to explain their reasoning in doing mathematics problems in almost every lesson. In contrast, less than one-half of discipline-oriented teachers did so. Beliefs also appear to have been related to what was taught (or to some underlying common factor or factors). For example, a smaller percentage of procedure-oriented teachers taught geometry topics and algebra than the other types of teachers. All (100 percent) of these teachers taught common and decimal fractions. A comparatively higher percentage of discipline-oriented and procedure-oriented teachers taught more measurement and related topics. Process-oriented teachers tended more often to teach topics in simple statistics and data analysis. These findings are illustrative rather than exhaustive. A study devoted to a more complete investigation of these data is currently underway.

These beliefs thus appear to have direct consequences for instructional practices as well as instructional content (although additional analyses are needed to explore these issues more thoroughly). Further, these beliefs seem to fall into fairly distinct categories which contain large proportions of teachers. In this context, even 10 percent of a teacher sample suggests that the characteristics shared by those in a group that large are not just coincidental but the product of systemic factors affecting training, responsibilities and teaching experiences. We (the authors) believe in coincidence. We also believe in probability. When a possible coincidence is also a low probability event, it can still happen as a coincidence but it is more logical to assume that the low probability event did not happen and that what has occurred is not merely a coincidence. Thus, it is more logical to assume that these commonalities of beliefs among US teachers are the consequences of systemic factors rather than a purely coincidental confluence of personal beliefs.

Given this context, the fact that US teachers, especially mathematics teachers, fell into several distinct categories, each with a non-negligible proportion of teachers, tells an interesting story. This says that the orientation to subject matter and other teacher beliefs important to instructional practice are likely the consequences of systemic factors. However, it also says that US education systems vary so much that, while they have commonalities sufficient to characterize teachers by a small number of categories or types, they do not have enough commonalities to allow them to be characterized by one or two types.

Other Aspects of Pedagogical Beliefs. We further examined teacher beliefs by sets of carefully structured survey questions. We presented each teacher with two real teaching situations (as real as could be constructed to fit several countries and quite real for US teaching). In mathematics these two situations involved introducing a specific new topic and responding to a type of student error in a different topic. In science, the situations dealt with different aspects of handling student misconceptions. In each case, teachers were asked to indicate and rank preferences for a series of given responses in that situation and were free to choose as many preferences as they desired.

The sets of responses that were provided had been systematically constructed to represent a range of pedagogical approaches that might stem from teacher beliefs including relying on group discussion, soliciting student views, relying on formal definitions of mathematical ideas, relying on technology, turning to a textbook as an authority, and others. The choices made were taken as an indication of the important factors of belief and characteristic pedagogy that underlay each teacher's specific pedagogical decisions. Teachers were also asked other, more direct questions concerning their beliefs about subject matter (mathematics or science), about their students, and about appropriate pedagogical strategies for teaching their subject matter.

The responses to these questions offered a more direct way to examine teachers' pedagogical beliefs. For these questions, we had the choices and priorities of each teacher for what they would do in response to each situation. Here we had data on 'enacted pedagogical beliefs', not so much what the beliefs underlying the pedagogical decisions were, but rather the teachers' indications of the decisions themselves. Conclusions about beliefs underlying decisions could only be inferred. However, broad commonalities in the decisions made would suggest practical commonalities of belief and provide more direct evidence that these differences were likely to affect instructional practices.

Exhibit 4.3 on page 90 shows one of the 'stimulus' pedagogical situations to which mathematics teachers for thirteen-year-olds were asked to respond. This particular situation relates to what teachers might do to introduce study of how to solve word problems that, although only exemplified rather than explicitly stated in the situation, hinge on proportionality. Teachers were given a set of specific responses to the situation, asked to indicate those they would use, and asked to rank those that they indicated. Each response related to a specific type of pedagogical action. The data displayed in Exhibit 4.3 summarizes the proportion of mathematics teachers who chose each response (regardless of the relative rank they gave it among their choices).

For this situation, almost all of the US seventh and eighth grade mathematics teachers (more than 80 percent) indicated that they would use proportional equations as their approach, or at least that it would be one of the approaches they would use. Almost the same percent of their French, Japanese, Czech, Korean, and Hong Kong counterparts indicated use of the same approach. However, almost the same high proportion (about 80 percent) of the Japanese teachers also chose an approach using graphing methods specific to the problem, the approach detailed in the textbook, or both.

Exhibit 4.3. *Percentage of Seventh- and Eighth-Grade Mathematics Teachers*
Selecting Various Approaches and the Mean Number of
Approaches Selected for Introducing How to Solve Word
Problems Involving Proportionality.

Each year many teachers must help their students learn to solve problems such as "Juan was
able to run 1.5 kilometers in 5 minutes. If he was able to keep up this same average speed,
how far would he run in 12.5 minutes?" If you needed to help your class learn how to solve
such problems, what approach do you believe would best help students learn?

	General Graphing	Proportional Equations	Textbook Strategy	Specific Graphing	Calculator Method	Small Groups	Mean # Approaches
USA	29.9 (2.8)	86.2 (2.3)	42.8 (3.5)	46.0 (3.3)	32.5 (2.7)	58.9 (2.9)	3.0 (0.1)
Australia	28.2 (2.6)	52.3 (3.1)	36.1 (3.1)	55.6 (2.7)	39.9 (3.2)	55.4 (3.1)	2.7 (0.1)
Belgium (Fl)	16.6 (2.6)	77.3 (2.6)	56.7 (3.7)	30.8 (3.5)	12.5 (2.0)	35.0 (3.3)	2.3 (0.1)
Czech Republic	19.3 (3.4)	82.0 (3.9)	55.1 (4.2)	52.2 (4.1)	24.7 (4.7)	40.9 (4.2)	2.7 (0.1)
France	28.0 (3.0)	81.8 (3.1)	25.3 (3.2)	67.8 (3.7)	27.0 (3.2)	49.0 (3.9)	2.8 (0.1)
Hong Kong	42.5 (4.1)	84.4 (3.1)	61.5 (4.8)	53.0 (3.9)	31.0 (4.2)	32.7 (4.0)	3.1 (0.1)
Hungary	18.2 (2.6)	34.5 (3.5)	32.9 (3.3)	58.8 (3.4)	19.0 (2.6)	10.5 (2.2)	1.8 (0.1)
Israel	28.8 (8.5)	68.2 (8.2)	34.9 (7.5)	45.7 (8.5)	20.0 (6.6)	55.1 (8.4)	2.6 (0.2)
Japan	54.4 (3.1)	83.8 (2.4)	78.2 (2.7)	79.2 (2.2)	34.2 (3.2)	62.7 (3.2)	3.9 (0.1)
Korea	50.2 (3.1)	84.0 (2.5)	42.9 (2.9)	64.0 (3.3)	26.2 (2.9)	50.3 (3.5)	3.2 (0.1)
Netherlands	29.7 (4.3)	37.9 (3.6)	77.4 (3.1)	61.4 (4.4)	32.5 (3.7)	62.9 (4.0)	3.0 (0.1)
New Zealand	30.9 (2.9)	34.0 (3.0)	29.7 (2.7)	61.9 (3.3)	43.6 (3.1)	74.4 (3.0)	2.8 (0.1)
Singapore	40.0 (3.3)	85.9 (2.2)	79.1 (2.7)	50.0 (3.9)	32.6 (3.4)	48.5 (3.6)	3.4 (0.1)
Spain	31.0 (3.8)	76.7 (3.5)	31.3 (3.8)	54.0 (4.0)	17.3 (2.9)	52.2 (4.1)	2.7 (0.1)
Sweden	15.2 (2.0)	17.0 (2.3)	57.7 (2.9)	36.9 (2.9)	26.1 (2.1)	67.1 (2.7)	2.3 (0.1)
Thailand	30.7 (4.3)	59.0 (5.2)	42.1 (6.0)	43.4 (5.3)	6.9 (2.5)	56.7 (5.5)	2.5 (0.2)

Strikingly, almost 70 percent of the Japanese teachers indicated that they
would try four or more of the six approaches presented. In contrast an
approach of having small groups work on the problem was the only other
approach that was chosen by more than 50 percent of the US teachers. There
seem to be clear, characteristic differences between Japanese and US teachers
in these responses. Japanese teachers seemed more willing to try multiple
approaches to a problem, while US teachers seemed to focus on one or two
particular approaches. US teachers appeared to act more as if there were one
correct approach, while Japanese teachers seem to have acted more as if it
were important to explore diverse approaches to every problem.

This same phenomenon was apparent in responses to the other mathematics
item. Exhibit 4.4 presents comparable data concerning the approaches they
would use when confronted with a specific mathematical misunderstanding, in
this case the relation of ratios to fractions. About 70 percent of Japanese teachers chose three of the approaches. Again, only one approach was chosen by more
than 70 percent of US teachers. Using a 'concrete situation' was chosen by about
80 percent of the US teachers. The second approach they chose varied somewhat. Teachers from Korea, Singapore, and the Netherlands demonstrated a similar propensity to using multiple approaches as the Japanese teachers.

Exhibit 4.4. *Percent of Seventh- and Eighth-Grade Mathematics Teachers Selecting Various Approaches and the Mean Number of Approaches Selected for Dealing with a Misconception about Ratios and Fractions.*

Many students have trouble relating ratios to fractions when they are asked to relate part of a set of objects to the whole set. For example, when asked "There are 2 boys in a class for every 3 girls in the class. What fraction of the students are boys?" Many students would answer 2/3 rather than 2/5. If you were working with a class in which many students had this kind of misunderstanding, what approach do you believe would best help students learn?

	Textbook Review	Concrete Situation	Student Discussion	Example Situations and Calculators	Set Theory	Relate to Concept of Ratio	Mean # of Approaches Chosen
USA	43.8 (3.3)	81.8 (2.1)	57.5 (3.4)	54.2 (3.1)	22.6 (2.4)	57.1 (2.3)	3.2 (0.1)
Australia	26.6 (2.5)	67.3 (2.6)	56.9 (3.3)	38.6 (2.4)	12.5 (2.5)	57.4 (2.8)	2.6 (0.1)
Belgium (Fl)	26.0 (3.1)	72.8 (3.3)	54.7 (3.1)	26.9 (3.0)	26.9 (3.1)	39.2 (3.8)	2.5 (0.1)
Czech Republic	42.3 (4.6)	65.3 (5.3)	61.9 (4.4)	33.3 (5.4)	10.8 (2.3)	52.4 (4.8)	2.7 (0.1)
France	19.8 (2.6)	80.7 (3.1)	65.3 (3.5)	43.8 (3.6)	16.8 (2.6)	41.5 (3.8)	2.7 (0.1)
Hong Kong	54.6 (4.8)	65.5 (3.9)	56.0 (5.3)	49.2 (5.4)	23.8 (3.3)	62.7 (4.4)	3.1 (0.2)
Hungary	12.8 (2.3)	41.5 (3.4)	27.2 (3.0)	18.7 (2.7)	13.8 (2.4)	61.5 (3.7)	1.8 (0.1)
Israel	8.3 (5.2)	84.1 (6.3)	35.7 (9.7)	46.5 (8.8)	32.8 (6.8)	51.9 (7.9)	2.6 (0.2)
Japan	61.0 (3.3)	70.1 (2.6)	71.2 (3.2)	49.9 (3.5)	34.0 (3.1)	70.5 (2.6)	3.6 (0.1)
Korea	41.9 (3.4)	68.6 (3.1)	52.5 (3.3)	41.0 (3.2)	45.7 (3.4)	68.7 (3.3)	3.2 (0.1)
Netherlands	59.9 (4.0)	77.1 (3.8)	60.1 (4.2)	34.1 (3.5)	11.0 (2.6)	75.5 (4.1)	3.2 (0.1)
New Zealand	26.4 (2.9)	70.3 (2.9)	55.6 (3.3)	36.2 (2.7)	12.7 (2.2)	63.9 (3.1)	2.7 (0.1)
Singapore	71.1 (2.8)	68.9 (3.2)	61.6 (2.9)	56.1 (3.4)	23.2 (2.6)	78.7 (2.9)	3.6 (0.1)
Spain	30.9 (4.0)	67.4 (3.9)	46.4 (4.2)	44.9 (4.2)	15.9 (2.8)	39.3 (4.1)	2.5 (0.1)
Sweden	41.7 (2.7)	73.0 (2.5)	69.1 (2.5)	43.0 (2.7)	4.2 (1.1)	29.3 (2.6)	2.6 (0.1)
Thailand	43.6 (5.3)	39.1 (4.1)	52.1 (5.3)	39.9 (5.7)	15.4 (3.5)	64.8 (4.5)	2.6 (0.2)

Again, Japanese teachers appear to have characteristically chosen multiple approaches to the situation while US teachers chose a single or more limited set of approaches. The pattern suggested in these data is supported by video-taped studies of US and Japanese eighth grade mathematics classes.[1] Corroboration was also seen in the observations gathered by the SMSO project.[2] Japanese teachers were seen in both of these data sources to be more open to solving problems with different approaches, while US eighth grade mathematics teachers focused on arriving at a correct answer rather than developing alternative approaches. It appears that other than Korea, Singapore, and the Netherlands, most other TIMSS countries were more like the US than they were like Japan.

Teacher Specialization. What teachers experience, believe, and practice is also affected by whether they function as specialists in mathematics or science education. Some US education systems and schools make official use of specialist teachers. Others give differing assignments to teachers that make some into *de facto* mathematics or science teaching specialists. Either form of specialization is likely to result in different experiences and practices.

At seventh and eighth grades (that is, for thirteen-year-olds), there were moderate differences among TIMSS countries in the use of science or mathematics specialist teachers. In the US, as Exhibit 4.5 makes clear, about 30 percent of classroom teachers specialized in science or mathematics or a combination of the two, at least in terms of their actual teaching load. The corresponding proportion in the Russian Federation was about 50 percent, about 40 percent in France and Spain, but less than 20 percent in Canada and Norway. This refers only to specialized teaching loads, to acting as *de facto* specialists. It does not speak to official designation as specialists, nor does it imply specialized training or knowledge (or, conversely, the absence of such training and knowledge).

Exhibit 4.5. *Percentage of Seventh- and Eighth-Grade Mathematics and Science Teachers with Various Types of Teaching Responsibilities.*

	>75% Science	>75% Science & Math	>75% Math
USA	12.5 (1.1)	6.8 (1.0)	12.6 (0.9)
Australia	11.0 (0.4)	13.3 (1.1)	12.0 (0.5)
Canada	6.2 (0.5)	6.9 (1.0)	6.6 (0.7)
Czech Republic	10.7 (1.0)	13.6 (1.1)	7.9 (0.6)
England	13.9 (0.2)	1.0 (0.4)	10.3 (0.2)
France	11.2 (1.3)	15.1 (1.7)	15.8 (1.4)
Hong Kong	12.5 (0.9)	22.2 (1.0)	12.5 (0.9)
Hungary	12.6 (0.9)	13.9 (1.1)	10.8 (1.1)
Israel	16.7 (5.1)	21.9 (5.0)	16.9 (5.6)
Japan	11.2 (0.3)	14.7 (1.1)	12.3 (0.3)
Korea	9.0 (0.8)	20.3 (2.0)	9.3 (0.8)
Netherlands	15.0 (1.2)	9.5 (1.4)	8.0 (0.4)
New Zealand	12.9 (0.3)	10.4 (1.1)	11.8 (0.3)
Norway	1.6 (0.5)	11.7 (1.0)	1.4 (0.3)
Russian Federation	14.5 (0.4)	21.5 (1.7)	13.3 (0.8)
Singapore	14.2 (0.3)	4.0 (0.5)	13.0 (0.3)
Spain	10.3 (1.9)	17.1 (2.1)	10.9 (1.7)
Sweden	8.4 (1.0)	21.4 (1.2)	1.2 (0.5)
Switzerland	7.0 (1.1)	16.6 (1.6)	5.4 (1.1)
Thailand	16.4 (1.9)	8.9 (1.4)	18.2 (2.7)

Without considering specialized training or knowledge, US thirteen-year-olds' mathematics teachers did not look that different from those *de facto* specialists in other TIMSS countries, with the possible exception of the Russian Federation, Israel, Hong Kong, and France. It appears not to have been the concept of specialist teachers or the *de facto* scheduling of teachers to function as specialists that made US teachers different. Any differences would seem necessarily to be attributed to differences in training, experience, and the resulting 'knowledge base' of each teacher. Science teachers at this level followed a pattern similar to that of mathematics teachers, and likely differed from their TIMSS countries' counterparts in similar ways to those in which US mathematics teachers differed.

One concomitant of specialization, or in some schools an alternative to it, is officially sanctioned cooperation among teachers. Thus, with officially sanctioned cooperation and sharing of responsibilities, teachers more prepared for teaching mathematics might take leadership roles in helping other teachers whose strengths lie in other areas. In Germany, Japan, Spain, and Switzerland, about 75 percent of the schools that contain the two grades for thirteen-year-olds had official policies related to promoting collaboration among teachers. About 58 percent of US schools did for seventh and eighth grades. For third and fourth grades (nine-year-olds), the proportion of US schools with such policies dropped to about 42 percent.

TEACHERS' BELIEFS, APPROACHES AND SPECIALIZATION: A SUMMARY

We presented a variety of data related to teachers' beliefs, their pedagogical approaches, and the use of *de facto* specialization in teaching science and mathematics among various TIMSS countries. Among the conclusions we suggest are supported by these data are the following:

- About three-fourths of US mathematics teachers for thirteen-year-olds held a somewhat formal view of mathematics. Many (eclectic teachers, about 40 percent of those surveyed) combined this with an emphasis on the importance of real-world applications, while others (discipline-oriented teachers, about 10 percent) put markedly less emphasis on real-world applications.

- About one-fourth of US seventh and eighth grade teachers (process-oriented teachers) did not share a formal view of mathematics, but did focus heavily on real-world applications. This group of teachers was likely to have far different practices than the first two groups. The fact that such differing orientations to teaching mathematics exist in the US suggests the fragmentation of views.

- Seventh and eighth grade science teachers were predominantly eclectic. That is, they held a formal view of science, but also emphasized real-world applications and considered creativity important.

- US science and mathematics teachers differed among themselves in their views of the disciplines, but fell into broad categories that suggested that these differences might be the consequences of factors that had produced differing training and experiences. Science teachers varied somewhat less than mathematics teachers.

- US mathematics teachers differed significantly among themselves on how important mathematics was in the real world and how this should affect instruction. Science teachers showed a similar difference in this area. Overall, for both science and mathematics teachers, more US teachers strongly agreed with the notion that it was important to emphasis the real-world applications of their discipline than did their Japanese counterparts.

- Science teachers did not generally view themselves as textbook followers. Mathematics teachers were more divided in their beliefs on this matter.

- Japanese teachers varied more in the approaches they chose to a stated instructional problem than did US teachers. Japanese teachers seemed characteristically to have believed it best to try a variety of approaches, while US teachers seemed characteristically to have focused on a correct approach and on obtaining answers.

- US schools often used specialized instructors in both mathematics and science for thirteen-year-olds. US practice reflected much the same approach as other countries on how manpower was allocated, but there is no guarantee of similarities in the *de facto* specialists' knowledge and training in the US compared to that in other countries.

- Schools in other countries seemed more often than US schools to have official policies that encouraged collaboration among teachers in the same discipline. This may well have facilitated *de facto* specialization as a way to cope effectively with different strengths and weaknesses in a school's faculty.

THE SCHOOL ENVIRONMENT

Schools as institutions, and as the final allocators of resources to classrooms, likely affect what is produced by US educational systems. The consequences for attainments of US students individually and as a whole are shaped, in part, by our typical choices of how to 'do' school. These choices include how we staff our schools, what our school principals and administrators do, and even the very conception of school implied by these things. We examine a few of these factors here.

The middle grades – those that contain most thirteen-year-olds – revealed staffing patterns that show they were often staffed more like elementary schools than secondary schools, the latter being distinguished by consistent specialist assignments (mathematics teachers, chemistry or science teachers, English teachers, etc.). Another look at Exhibit 4.5 on page 92 is relevant here. This display shows that mathematics and science specialists were the exception, rather than the rule, at this level. The proportion of science 'specialists' (those with assignments that devoted more than 75 percent to teaching science or science and mathematics) ranged from about 13 percent to over 35 percent.

The proportion of mathematics 'specialists' (those with assignments that devoted more than 75 percent to teaching mathematics or mathematics and science) covered essentially the same range. There were virtually no major differences between the proportions of mathematics and science specialist teaching assignments in any of these countries. The US, however, was closer to the lower end of the range.

Overall, about one sixteenth of US seventh and eighth grade classroom teachers had assignments that were more than three-quarters devoted to teaching both mathematics and science. About one eighth had assignments more than three-quarters devoted to mathematics, with another eighth having assignments similarly devoted primarily to science. Altogether, just over 30 percent of such US seventh and eighth grade teachers had specialized assignments in mathematics and science, although as much as a fourth of their time could be devoted to teaching something other than mathematics or science. This suggests that a significant proportion of US students at this level received mathematics or science instruction from those who were not even *de facto* specialists in those disciplines.

Principals' Roles and Schools' Nature. A look at the roles and functions of principals in individual schools across several TIMSS countries provides an interesting insight into US schools. Exhibit 4.6 shows the time spent on different activities by principals of schools containing thirteen-year-olds. These can be grouped into four categories which include teaching, administrative functions, school relations, and professional development. These four categories account for the majority but not all of principals' time.

'Teaching' is self-explanatory. 'Administrative functions' include hiring teachers, performing internal administrative tasks, and counseling and disciplining students. 'School relations' includes representing the school in the community, representing the school at official meetings, talking with parents, and responding to requests from district, state, or national education officials. 'Professional development' includes giving demonstration lessons to teachers, discussing educational objectives with teachers, initiating curriculum revision, planning, or both, training teachers, and other professional development activities (including the principal's own professional development). These activities along with 'other activities' were the specific activities to which principals were asked to respond.

The role of principals – as indicated by the functions or activity categories on which they spend their time – varied considerably across countries. For principals of US schools containing seventh and eighth grades, school relations took over a third of their time and administrative functions almost as much. Professional development took another fourth of their time but teaching, on average, took less than two percent. The rest was devoted to other activities. Of the 93 percent of US principals' time accounted for here, less than 30 per-

Exhibit 4.6. *Mean Number of Hours per Month Principals Report Spending on Each Activity.*

	Counseling and disciplining students	Discussing educational objectives with teachers	Giving a demonstration lesson	Hiring teachers	Initiating curriculum revision and/or planning	Internal administrative tasks	Other activities
USA	24.6 (1.7)	12.7 (0.9)	0.8 (0.2)	2.9 (0.3)	11.1 (0.9)	22.9 (1.5)	11.2 (1.2)
Australia	14.0 (1.5)	14.9 (1.8)	0.1 (0.1)	3.5 (0.6)	13.8 (1.9)	46.6 (3.0)	17.5 (2.3)
Belgium (Fl)	16.3 (2.1)	9.2 (0.9)	0.0	2.8 (0.3)	5.5 (0.6)	37.2 (4.3)	21.0 (2.6)
Canada	20.8 (1.6)	8.8 (0.8)	0.6 (0.1)	1.8 (0.2)	7.0 (0.7)	37.9 (3.2)	9.8 (1.1)
Czech Republic	9.0 (0.6)	9.7 (0.8)	0.7 (0.2)	2.4 (0.2)	6.5 (0.5)	46.7 (2.3)	17.9 (1.6)
France	16.6 (1.5)	10.9 (1.0)	0.0	0.5 (0.1)	5.1 (0.7)	40.7 (3.1)	14.3 (1.5)
Germany	11.6 (0.9)	11.5 (0.9)	0.1 (0.1)	0.7 (0.2)	5.0 (0.7)	41.0 (3.0)	9.6 (1.3)
Hong Kong	8.0 (1.3)	8.1 (1.0)	0.2 (0.1)	4.1 (0.6)	7.8 (1.1)	65.1 (5.3)	10.1 (1.7)
Hungary	12.2 (0.8)	18.8 (1.0)	0.8 (0.2)	3.6 (0.3)	7.5 (0.6)	28.9 (1.9)	20.1 (1.5)
Israel	12.9 (1.5)	13.7 (2.5)	0.8 (0.3)	3.9 (0.8)	9.9 (1.5)	27.4 (2.8)	13.3 (2.3)
Japan	6.0 (0.9)	6.8 (0.8)	0.1 (0.0)	5.5 (0.7)	8.9 (0.8)	37.7 (2.8)	22.1 (2.0)
Korea	9.2 (1.2)	7.4 (1.0)	1.9 (0.9)	1.6 (0.5)	6.3 (1.0)	39.6 (2.7)	14.3 (1.7)
Netherlands	12.2 (1.6)	11.1 (1.4)	0.1 (0.1)	1.8 (0.4)	5.9 (1.0)	33.4 (4.3)	43.9 (8.3)
New Zealand	14.6 (1.2)	13.4 (0.9)	0.3 (0.1)	4.4 (0.3)	12.7 (1.0)	65.5 (4.6)	19.9 (2.4)
Norway	8.9 (0.8)	7.2 (0.6)	1.3 (0.2)	2.4 (0.2)	5.3 (0.5)	38.2 (2.5)	18.7 (2.3)
Russian Federation	17.7 (1.1)	17.1 (1.2)	2.0 (0.3)	2.4 (0.3)	9.4 (0.8)	31.3 (1.8)	16.0 (1.1)
Singapore	22.2 (1.9)	12.1 (1.3)	0.6 (0.2)	1.3 (0.3)	11.1 (1.1)	35.4 (2.8)	16.7 (2.3)
Spain	5.9 (0.6)	5.7 (0.4)	1.0 (0.5)	0.5 (0.1)	5.2 (0.4)	21.4 (1.5)	7.7 (1.0)
Sweden	8.7 (0.8)	10.4 (0.9)	0.0	2.3 (0.3)	13.1 (1.1)	37.8 (3.0)	23.1 (1.9)
Switzerland	4.4 (0.5)	4.3 (0.4)	0.0	2.2 (0.3)	3.7 (0.3)	24.2 (2.6)	15.7 (1.3)
Thailand	13.9 (1.6)	11.8 (1.4)	2.8 (1.2)	1.5 (0.4)	9.9 (1.1)	18.6 (2.3)	11.4 (1.6)

	Professional development activities	Representing the school at official meetings	Representing the school in the community	Responding to requests from district, state, or national education officials	Talking with parents	Teaching	Training teachers
USA	8.3 (0.8)	11.9 (0.7)	9.0 (0.8)	12.9 (1.0)	22.2 (1.5)	2.4 (0.6)	7.1 (0.9)
Australia	8.7 (0.7)	15.3 (1.3)	10.1 (0.9)	15.4 (1.6)	18.6 (1.4)	4.4 (0.8)	2.4 (0.4)
Belgium (Fl)	6.3 (0.5)	7.8 (0.9)	7.0 (0.8)	5.9 (0.5)	7.0 (0.8)	0.1 (0.0)	3.0 (0.3)
Canada	6.2 (0.6)	8.5 (0.5)	5.0 (0.5)	7.5 (0.6)	14.6 (0.9)	5.3 (0.8)	4.0 (0.5)
Czech Republic	14.9 (1.0)	7.2 (0.6)	6.4 (0.7)	15.0 (1.3)	8.5 (0.6)	34.2 (1.2)	2.7 (0.3)
France	3.2 (0.4)	8.6 (1.0)	4.6 (0.4)	21.7 (2.0)	15.2 (1.0)	3.5 (0.9)	1.5 (0.3)
Germany	6.5 (0.8)	3.9 (0.4)	2.9 (0.4)	7.4 (0.6)	12.1 (1.1)	49.0 (3.1)	5.1 (0.6)
Hong Kong	6.8 (1.0)	6.4 (0.8)	7.4 (1.4)	5.9 (0.6)	5.5 (0.9)	4.0 (1.1)	4.2 (0.7)
Hungary	11.7 (0.7)	8.2 (0.4)	9.0 (0.5)	6.5 (0.7)	11.6 (0.8)	35.2 (1.6)	6.7 (0.7)
Israel	7.8 (1.0)	8.3 (0.7)	5.7 (0.7)	10.5 (1.4)	12.6 (1.3)	26.6 (3.6)	6.3 (0.9)
Japan	11.9 (1.0)	19.7 (1.3)	7.7 (0.6)	5.2 (0.4)	7.5 (0.9)	0.9 (0.4)	6.2 (0.4)
Korea	12.8 (1.5)	7.8 (1.0)	5.6 (1.0)	5.2 (0.8)	8.2 (1.0)	3.1 (0.8)	6.4 (1.0)
Netherlands	8.8 (1.0)	3.1 (0.4)	6.1 (0.8)	1.7 (0.2)	6.9 (0.8)	14.1 (3.4)	8.0 (1.4)
New Zealand	10.0 (0.7)	14.8 (0.9)	13.1 (0.9)	13.6 (0.9)	18.7 (1.2)	9.0 (1.2)	3.9 (0.4)
Norway	6.0 (0.6)	7.5 (0.6)	3.3 (0.3)	13.8 (1.1)	7.0 (0.5)	23.2 (1.5)	7.7 (0.7)
Russian Federation	20.3 (1.4)	8.6 (0.5)	6.7 (0.4)	8.1 (0.8)	15.9 (1.3)	40.2 (1.6)	8.9 (0.9)
Singapore	9.4 (1.0)	7.7 (0.5)	2.7 (0.2)	9.7 (1.0)	16.6 (1.5)	3.8 (0.6)	6.6 (1.0)
Spain	5.7 (0.5)	3.8 (0.5)	4.4 (0.6)	8.1 (0.6)	9.2 (0.6)	26.3 (2.0)	1.5 (0.3)
Sweden	7.2 (0.7)	4.1 (0.4)	8.8 (0.9)	7.0 (0.6)	8.5 (0.6)	4.3 (0.8)	2.6 (0.3)
Switzerland	4.2 (0.3)	6.2 (0.5)	2.2 (0.3)	4.0 (0.3)	4.3 (0.4)	87.0 (5.4)	1.0 (0.2)
Thailand	11.0 (1.2)	18.9 (1.7)	13.7 (1.3)	14.1 (1.4)	16.7 (2.0)	3.9 (1.0)	9.3 (1.3)

cent of that time was devoted to direct educational activities while almost 75 percent of their time was devoted to more managerial and public relation functions.

In many TIMSS countries principals or headmasters had as their primary roles teaching and administrative tasks (in the narrower sense asked directly – related to school functioning – and not in the more inclusive 'administrative function' identified earlier). These two took about half of the time of these principals in the Czech Republic, Germany, Hong Kong, Norway, Spain, and Switzerland. In Japan, they accounted for about 26 percent of the principals' time and in the US less than 20 percent. Principals essentially did not teach in Belgium (Flemish), France, Hong Kong, Japan, Korea, Singapore, Sweden, Thailand, and the US.

US principals in schools containing seventh and eighth grades likely viewed their mission as more managerial and administrative and less as directly involved in educational functions (teaching and aiding in the professional development of teachers). Certainly this was how they spent their time.

What does this conception by educational systems of the role of principals possibly imply about how those same systems see schools and schooling? In many countries, the relative time allocations suggest that the educational systems saw schools as teaching institutions and principals as both the head of the organization and an essential part of providing that teaching. In the US, school appears to have been seen first as an organization and only second as an organization whose product was education.

It may well be an implication of how principals' roles are conceived, and the expectations that underlie them, that schooling is not, for them, about teaching. This is not meant to imply fault on the part of principals or to imply that they do not work hard at the tasks expected of them: counseling and disciplining students, community and parent relations, and organizational relations. In fact, adding up the 26 percent of time US principals at this level spent on hiring and training teachers, curriculum-related issues, and professional development to the 16 percent spent on direct administrative tasks and teaching, still only accounts for less than half of their time.

This seems an indication that resources were diverted from the core functions of schools. Principals' time did not seem to be devoted directly to functions that supported learning and teaching tasks, but rather to functions that supported the organization and its internal and external relations. If we are to again think of schools as support organizations for teaching and learning, then we may well have to challenge the implicit assumptions of what schools should be and the societal models that have fed our conceptions of what is to be expected of schools and those who work within them. This factor alone does not explain achievement differences across countries as is true with almost any of the factors we have discussed in the preceding chapters. However, when taken together to define the system we have, this factor seems likely to be important. If the system allows the instructional content to be determined by teachers and schools, then the fact that principals' time is so consumed by other elements may be significant. This is discussed in greater detail later. The data in the exhibit above show, cross-nationally, remarkably different approaches to several functions that we in the US often associate with schools.

School organization. The sets of grades distributed to various levels of schools were organized differently in the TIMSS countries. This is particularly relevant to where the middle grades – the grades containing thirteen-year-olds – were situated. Exhibit 4.7 shows the distribution of US schools that contained third and fourth grades into categories according to the ranges of school grade levels that were covered by those schools.

The pressing question to answer through Exhibit 4.7 is, 'What kinds of grade ranges were typical of the schools that contain third and fourth grades?' Third and fourth grades were typically located in schools with ranges from pre-kindergarten or kindergarten to fifth or sixth grade. Almost 60 percent of the schools that contained third and fourth grades had those grade ranges.

Exhibit 4.8 displays the corresponding distribution of school types for seventh and eighth grades in the US and several other countries. Singapore, the Russian Federation, Belgium (Flemish), Australia, England, New Zealand, the Netherlands, Thailand, and Germany, had most of their children in systems that go up into the secondary. Korea and France were organized more like the US. Hungary, Hong Kong, and the Czech Republic tended more towards a first through eighth grade organization. Since Korea typically did well on the TIMSS test, this factor of how the middle grades were placed must be thought of as part of the defining elements of the system and not as a single explanation for lower US achievements. For example, we must keep in mind differences in the organization of curricula in Korea. It may be that grouping the middle grades with lower grades becomes more problematic when curricula are not clearly articulated and where the personal preferences and comfort of the teachers with subject matter determine the content of instruction. Considering the middle grades as the culmination of elementary schooling allows decision-makers not to push forward with new material more typical of secondary schooling, and frees more time for what they considered as necessary review.

In what kinds of schools did we most often find seventh and eighth grades? The three school types (based solely on grade range) with the largest proportions of US seventh and eighth graders were sixth to eighth grades, seventh and eighth grades, and kindergarten to eighth grade (47, 18, and 9 percent of the schools, respectively). Clearly US seventh and eighth grades were grouped with lower grades. Almost one-fifth of the schools with these two grades contained these two grades alone.

None of the three most dominant school types contained grades higher than eighth and two contained grades lower than eighth. Essentially, about 75 percent of the US schools that contained seventh and eighth grades contained no grades higher than eighth grade and about 56 percent contained lower grades. Clearly, in the US, seventh and eighth grades were situated in isolated 'seventh and eighth only' schools and more typically in elementary schools or middle schools that culminated with eighth grade.

None of the three school types most prevalent in the US were even represented among the German school types (again based solely on grade ranges). The three most prevalent German school types were fifth through tenth grades, fifth through thirteenth grades, and seventh through tenth grades (about 31, 24, and 13 percent, respectively). These three most-prevalent types accounted for about 68 percent of German schools that contained seventh and eighth grades. None of these ranges were represented among US school types.

For each of these three most prevalent German school types, seventh and eighth grades were either situated in the middle of the grades covered or seventh was the lowest grade covered. In each of these three cases, at least ninth and tenth grades were included in any school that contained seventh and eighth grades. Seventh and eighth grades in Germany seem clearly to have been situated in either lower secondary or more inclusive schools.

Why this concern for where seventh and eighth grades were situated? Elementary and secondary, even 'junior secondary' schools had definite connotations in most TIMSS countries. In particular, the type of school has traditionally been tied to the kinds of expectations for its students' accomplishments, the experience and training of the staff assigned to it, and the kinds of curricula that might be considered appropriate.

Elementary schools have typically been viewed as filled with developmentally younger students for whom more complex performances in any discipline – in this case, mathematics and the sciences – would be expecting them to accomplish things for which they might not be developmentally ready. Teachers have been assigned to these schools accordingly. In the US, they have typically had a generalist's training in the several areas (language arts, mathematics, science, social studies, etc.) for which they would be responsible with a premium placed on nurturing and fostering the development of students rather than refining and extending subject matter knowledge. With these constraints in place, curricular goals have been shaped with more limited expectations and demands on both students and teachers.

Such limits would be unlikely for these grades in countries such as Australia, Hong Kong, Korea, Singapore, Belgium (Flemish), the Netherlands, and Thailand in which the grades were situated in secondary school with raised expectations of students, teachers with more specialized training and experience in mathematics and science, and challenging curricula linked much more

Exhibit 4.7. *Organization of School Grade Levels in Schools that Contained Third and Fourth Grades (percent of schools with various grade level ranges).*

	USA	Australia	Canada	Czech Republic	England	Hong Kong	Hungary	Israel
pk-13			0.2 (0.2)					
pk-12	2.9 (2.6)				1.4 (1.0)			
pk-10					0.8 (0.8)			
pk-9	0.9 (0.9)		2.3 (1.9)					
pk-8	8.1 (2.9)		15.3 (1.8)					
pk-7			0.5 (0.3)					
pk-6 & 8	0.3 (0.3)							
pk-6	8.5 (2.9)		10.5 (5.1)					
pk-5	12.5 (2.6)		2.7 (1.0)			54.6 (2.4)		
pk-4	2.5 (1.4)		0.4 (0.3)			1.7 (1.7)		
pk-k & 4-8			0.5 (0.5)					
pk-3 & 6								
pk & 2-7								1.6 (1.6)
pk & 1-6			16.2 (2.0)					
pk & 1-4								
k-13		0.2 (0.2)						
k-12	1.3 (1.2)	1.6 (1.5)	1.1 (0.4)					
k-11			0.4 (0.4)					
k-10		2.0 (0.8)					0.8 (0.8)	
k-9			1.4 (0.5)	0.9 (0.9)				
k-8	3.2 (1.6)		9.0 (2.3)	2.9 (1.3)			1.1 (0.8)	2.1 (2.1)
k-7		26.5 (2.9)	9.7 (1.9)					
k-6	15.0 (3.4)	53.2 (3.1)	18.2 (2.2)					7.7 (3.7)
k-5	22.8 (4.4)		3.2 (1.0)					
k-4 and 6-9			0.1 (0.1)					
k-4	7.8 (3.1)		1.0 (0.6)	2.4 (1.1)				
k-3								
k-1 & 3-5								
k-1 & 3-4				0.3 (0.3)				
1-13							0.4 (0.4)	
1-12		0.6 (0.6)	0.1 (0.1)					1.3 (1.3)
1-11		0.2 (0.2)					0.9 (0.6)	
1-10							5.0 (1.9)	
1-9			0.1 (0.1)	22.9 (3.6)			1.1 (0.8)	
1-8	1.0 (1.0)		0.9 (0.5)	52.3 (4.3)			90.3 (2.5)	11.8 (4.6)
1-7		9.6 (2.9)		1.4 (1.0)				2.6 (2.6)
1-6 &13								1.9 (1.9)
1-6 & 8								0.8 (0.8)
1-6	1.1 (0.8)	3.1 (1.8)	2.0 (1.3)	0.7 (0.7)		92.4 (2.1)		66.5 (7.2)
1-5	0.3 (0.3)		0.5 (0.3)	1.3 (0.9)		0.5 (0.5)		1.9 (1.9)
1-4				12.3 (3.0)				
1-3								
1 & 3-6							0.6 (0.7)	
1 & 3 & 5-6							0.6 (0.6)	
2-7					0.6 (0.6)			1.8 (1.9)
2-6	0.5 (0.5)							
2-5	1.1 (1.1)				24.2 (3.2)			
2-3								
2 & 4 & 6							0.7 (0.7)	
3-8			0.2 (0.2)	0.8 (0.8)				
3-7		1.4 (0.6)		0.9 (0.9)				
3-6			1.4 (1.3)			6.0 (2.1)		
3-5	4.1 (1.8)					0.8 (0.8)		
3-4			0.7 (0.5)				5.2 (1.7)	
4-8	2.7 (2.6)		1.1 (0.7)					
4-7			0.1 (0.1)		7.7 (1.6)			
4-6	2.7 (2.7)	0.3 (0.3)	0.3 (0.2)					
5-12		1.2 (1.2)						
6-12					2.2 (1.3)			
only 4	0.7 (0.7)						0.4 (0.4)	

Exhibit 4.7. (Continued).

Japan	Korea	Netherlands	New Zealand	Norway	Singapore	Thailand	
							pk-13
							pk-12
							pk-10
							pk-9
							pk-8
							pk-7
							pk-6 & 8
	3.4 (1.5)					9.3 (2.5)	pk-6
							pk-5
							pk-4
							pk-k & 4-8
						0.1 (0.1)	pk-3 & 6
							pk & 2-7
						2.8 (2.6)	pk & 1-6
						0.1 (0.1)	pk & 1-4
							k-13
							k-12
							k-11
							k-10
				6.6 (2.3)		14.2 (4.3)	k-9
						0.5 (0.5)	k-8
						0.7 (0.5)	k-7
	46.2 (3.0)	100.0		39.2 (4.1)		60.5 (5.9)	k-6
	0.6 (0.6)			2.3 (1.6)		0.2 (0.2)	k-5
							k-4 and 6-9
				1.0 (1.0)		0.4 (0.4)	k-4
				1.3 (1.3)			k-3
						0.1 (0.1)	k-1 & 3-5
							k-1 & 3-4
			2.2 (0.9)				1-13
			1.1 (1.1)				1-12
					0.2 (0.2)		1-11
					1.9 (1.1)		1-10
			0.9 (0.9)	5.3 (2.0)		5.3 (4.8)	1-9
			36.4 (3.3)				1-8
			0.8 (0.8)		0.6 (0.6)		1-7
							1-6 &13
							1-6 & 8
100.0	49.1 (3.1)		58.6 (3.6)	39.1 (4.6)	97.3 (1.3)	5.5 (2.0)	1-6
	0.7 (0.7)			1.7 (1.2)			1-5
						0.3 (0.3)	1-4
				1.9 (1.4)			1-3
				0.9 (0.9)			1 & 3-6
							1 & 3 & 5-6
							2-7
							2-6
							2-5
				0.7 (0.7)			2-3
							2 & 4 & 6
							3-8
							3-7
							3-6
							3-5
							3-4
							4-8
							4-7
							4-6
							5-12
							6-12
							only 4

Exhibit 4.8. *Organization of School Grade Levels in Schools that Contained Seventh and Eighth Grades (percent of schools with various grade level ranges).*

	USA	Australia	Belgium (Fl)	Canada	Czech Republic	England	France	Germany	Hong Kong	Hungary	Israel
pk-13								1.2 (1.2)			
pk-12	0.6 (0.6)			0.2 (0.2)		4.5 (1.4)					
pk-11											
pk-9				1.0 (0.9)							
pk-8	3.8 (1.9)			16.1 (1.9)							
pk-7				1.0 (0.6)							
pk-6 & 8											
pk-k & 4-8				0.4 (0.4)							
pk-k & 2-8					1.2 (1.2)						
k-13		2.2 (1.5)									
k-12		6.1 (1.6)		1.2 (0.5)							
k-11											
k-10		0.7 (0.5)								0.7 (0.7)	
k-9				4.3 (2.6)							3.9 (3.9)
k-8	8.9 (3.0)			9.2 (1.8)	4.0 (2.3)					1.4 (1.0)	
k-7											
k-6 & 8											
k-3 & 7-9											
k-3 & 5-11											
k & 6-9											
1-13		1.0 (0.6)								0.7 (0.7)	
1-12		1.8 (1.2)		0.2 (0.2)			1.0 (1.0)				2.7 (2.7)
1-11										2.1 (1.2)	
1-10								2.9 (1.5)		5.5 (2.0)	
1-9 & 11-13					0.9 (0.9)						
1-9				0.3 (0.2)	28.6 (5.3)			3.4 (2.5)		1.4 (1.0)	3.1 (3.0)
1-8				0.6 (0.4)	60.1 (5.3)					88.3 (2.8)	16.1 (5.3)
1-7											
1-6 & 8-9											
1-5											
1-4 & 7-9								0.7 (0.7)			
1-3 & 5-11											
1-3 & 5-10											
1-3 & 5-9											
1-2 & 4-5											
2-12						0.9 (0.9)					
3-8				0.2 (0.2)	1.1 (0.8)						
4-12				0.7 (0.5)		0.9 (0.9)					
4-9				0.5 (0.5)							
4-8				1.3 (0.7)							
4 & 7-9											
5-13								24.3 (3.4)			
5-12		1.8 (1.1)				1.1 (1.1)	0.9 (0.9)	10.2 (3.8)			
5-11				0.0							
5-10								30.9 (4.9)			
5-9								10.5 (1.5)			
5-8	5.1 (1.9)			0.9 (0.4)							
6-13		0.3 (0.3)				1.9 (1.1)		2.2 (1.5)			
6-12	4.1 (2.6)						42.8 (3.4)	4.4 (2.0)			
6-11							27.9 (3.4)	0.7 (0.7)			
6-9		0.5 (0.5)		0.8 (0.7)			5.0 (2.1)	83.0 (3.1)			
6-8 & 10-13					0.9 (0.9)						
6-8	46.8 (5.1)			4.4 (0.9)			1.7 (1.2)				3.9 (3.8)
6-7											4.1 (3.8)
7-13		2.2 (1.5)	36.6 (4.8)	0.1 (0.1)		1.3 (1.2)		3.6 (2.6)	88.1 (4.2)		
7-12	5.9 (2.1)	44.9 (3.5)	30.3 (3.4)	2.2 (0.6)			3.3 (2.0)				22.4 (8.4)
7-11				15.5 (2.4)			0.7 (0.7)				3.4 (3.4)
7-10 & 12-13											
7-10 & 12		0.9 (0.9)									
7-10		8.0 (2.5)	4.2 (1.8)	2.0 (2.0)				13.5 (2.9)	9.4 (3.6)		
7-9	4.2 (1.8)			18.3 (3.2)			0.8 (0.8)				36.7 (8.3)
7-8 & 10-11									2.6 (2.6)		
7-8 & 10-12											
7-8	18.5 (3.9)		28.9 (4.5)	8.6 (2.0)							
8-13		5.2 (1.1)									
8-12	0.5 (0.5)	24.2 (2.4)		6.6 (1.0)			11.2 (0.5)	2.8 (2.3)			
8-11		0.2 (0.2)						1.5 (1.5)			
8-10				3.5 (1.9)							
8-9	0.7 (0.7)							1.5 (1.5)			
9-13											
9-11											3.7 (3.7)
10-12								0.1 (0.1)			
only 7	0.9 (0.9)										

Exhibit 4.8. *(Continued).*

Korea	Netherlands	New Zealand	Norway	Russian Federation	Singapore	Spain	Sweden	Switzerland	Thailand	
										pk- 13
0.6 (0.6)										pk- 12
						6.1 (1.8)				pk- 11
						2.8 (1.4)		0.2 (0.2)		pk- 9
										pk- 8
						37.8 (3.8)				pk- 7
						0.3 (0.2)				pk-6 & 8
						1.1 (1.1)				pk-k & 4-8
										pk-k & 2-8
										k- 13
						5.4 (2.0)				k- 12
				11.6 (3.5)		2.3 (1.4)				k- 11
				0.5 (0.5)						k- 10
			7.7 (2.6)				14.3 (3.6)			k- 9
						37.8 (3.8)				k- 8
						1.9 (1.1)				k-7
						0.5 (0.5)				k-6 & 8
							1.0 (1.0)			k-3 & 7-9
				1.3 (0.8)						k-3 & 5-11
							0.8 (0.8)			k & 6-9
		3.5 (0.8)								1-13
										1-12
						1.1 (1.1)				1-11
				61.2 (4.5)	0.3 (0.3)					1-10
				1.6 (1.4)	2.8 (1.6)			5.0 (2.0)		1-9 & 11-13
			17.6 (2.7)	2.2 (1.1)			9.1 (2.8)	23.1 (3.8)		1-9
								0.2 (0.1)		1-8
						2.9 (1.4)		0.2 (0.2)		1-7
			1.1 (0.8)					0.4 (0.3)		1-6 & 8-9
										1-5
										1-4 & 7-9
				19.4 (3.8)						1-3 & 5-11
				0.5 (0.5)						1-3 & 5-10
				0.9 (0.5)						1-3 & 5-9
										1-2 & 4-5
										2-12
										3-8
										4-12
							4.9 (1.6)	1.3 (1.3)		4-9
										4-8
							0.8 (0.8)			4 & 7-9
										5-13
										5-12
				0.8 (0.7)						5-11
								0.1 (0.1)		5-10
							1.1 (1.0)	3.1 (1.5)		5-9
										5-8
										6-13
								0.8 (0.8)	1.0 (1.0)	6-12
										6-11
							1.6 (1.1)	18.3 (2.1)		6-9
										6-8 & 10-13
										6-8
										6-7
		12.4 (1.6)			4.0 (1.5)			2.2 (0.7)		7-13
1.9 (1.1)	59.6 (3.6)						1.0 (1.0)	1.3 (0.9)	72.3 (4.1)	7-12
	0.2 (0.2)			74.5 (3.8)				0.8 (0.5)	4.0 (1.5)	7-11
				0.9 (0.9)						7-10 & 12-13
										7-10 & 12
	40.2 (3.6)				15.0 (3.5)			4.6 (1.9)	5.7 (1.9)	7-10
94.1 (2.5)			72.3 (2.2)			62.5 (5.2)	38.4 (3.2)	13.1 (2.5)		7-9
							2.9 (2.7)			7-8 & 10-11
										7-8 & 10-12
2.7 (1.9)					2.5 (1.4)			2.3 (1.1)		7-8
		0.8 (0.8)								8-13
								1.5 (1.5)		8-12
										8-11
0.6 (0.6)										8-10
										8-9
		83.3 (1.6)								9-13
										9-11
										10-12
			1.3 (1.0)							only 7

directly to what would be studied in later secondary school grades. However, this is not, as mentioned before, a 'magic bullet' or sole explanation for achievement since some top achieving countries (such as the Czech Republic) have an organization not too dissimilar from that of the US. However, this factor certainly seems to be another aspect of an explanatory 'mosaic.'

THE SCHOOL ENVIRONMENT: A SUMMARY

Our examination of the school environment suggests several important ways in which US schools differed from those in other countries in how they were organized and conceived, and differed in ways which, when combined with factors discussed in other sections, are likely to affect student attainments in mathematics and the sciences. Some important points are the following:

- About one-third of US seventh and eighth grade mathematics and science instruction was provided by teachers that had three quarters or more of their teaching assignment devoted to science or mathematics, or both. Many other TIMSS countries showed similar patterns.

- US principals of schools that contained seventh and eighth grades spent three times the amount of time on internal and external relations as they did on teaching and internal administration. Even when other activities such as training and giving professional development to teachers were added, these US principals still spent less time on the educational aspects of their school than on its institutional aspects.

- This use of time likely sent messages to other school personnel as to what 'school' was for, that is, that it was primarily an institution managing people and relations rather than a place for teaching and learning.

- In the absence of clearly defined curricular intentions, where classes for thirteen-year-olds were located among school types likely had implications for the training and knowledge of teachers assigned there, for what was expected of students, and for what was regarded as reasonable curricular goals in science and mathematics at that level. Typically, the seventh and eighth grades in the US were isolated from secondary schools. Most teachers with majors in mathematics or the sciences were likely located in secondary schools.

STUDENT FACTORS

We have examined many school-related factors that affect educational quality. Other factors may complicate untangling the interrelated affects of schooling. For example, things outside our formal educational systems affect what students accomplish, even when given access to similar educational possibilities. These include individual student interest and motivation, as well as how they use their time both inside and outside of school.

Student-centered problems. The US had a high percentage of students attending schools in which tardiness and absenteeism occurred regularly. Exhibit 4.9 presents information on how often school staff reported having to deal with eighth grade students who arrived late at school or who were absent (without acceptable reasons or permission). The exhibit displays whether school staffs had to deal with these behaviors daily and weekly (one or more, but not all days of the week). Even though the US percentages of schools that had to deal with these problems daily were as high, or higher, than those of other countries, the number of students involved in both cases on a given day was less than four percent. This four percent did not necessarily represent the same students from day to day. At this grade level, the frequency of tardiness and of absenteeism truly represents an American problem, but one that is not unique to US schools.

Exhibit 4.9. *Percent of Schools* Reporting Weekly or Daily Tardiness or Absenteeism for Eighth Grade Students.*

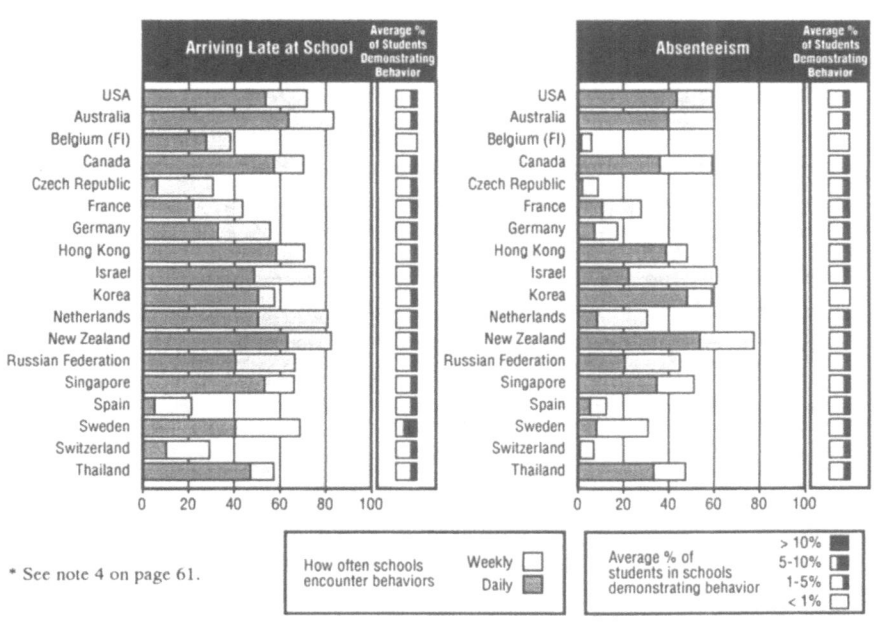

* See note 4 on page 61.

These phenomena have powerful consequences for US education systems' planning for providing access to learning and sequences of learning experiences. Even if a system and its schools provide access to educational possibilities, individual students can not have access to those possibilities if they are late or absent. Less directly, these recurrent problems affect the learning climate of classrooms likely lowering at least some students' (other than those habitually late or absent) perceptions of the importance of the opportunities provided within those classrooms. The schools for which data was reported in Exhibit 4.9 seem likely to have suffered both the direct and indirect consequences of these problems.

Other problems such as students creating classroom disturbances, cheating, or using tobacco, alcohol, or illegal drugs were no more pronounced in the US than in other TIMSS countries. The same was true of students possessing weapons and of inappropriate sexual behavior. Most of these problems were in schools attended by only a small proportion of US students and only a small percentage (one or two percent) of students within those schools were involved in such behaviors each day. There was a somewhat higher percentage of schools in which profanity had to be dealt with by school staff on a weekly or daily basis.

Student beliefs and attitudes. American eighth grade students believed that most of them did well in both mathematics and science (about 85 percent). They reported overwhelmingly (more than 70 percent) that they liked both subjects and most (more than 70 percent) said they enjoyed learning mathematics and sciences. Were these reports to be taken at face value, there should have been little problems with the science and mathematics achievements of these students. Obviously, that was not true.

Several explanations present themselves for these discrepancies between beliefs and performance. First, it might be that students were simply overly optimistic, especially about their own performances. If so, this would suggest that student attitudes presented no barriers to science and mathematics achievement, at least no more so than in other countries.

Alternatively, the US students may simply have responded to these questions with the kinds of responses that they thought those responsible for the study, or those in authority in their schools, would want to hear. A third possibility is that the US students likely responded only to the school mathematics and science that they had experienced.

The second or third possibilities are supported by further data on US student attitudes in eighth grades. About 40 percent of these same students indicated that mathematics and science were boring. When asked if mathematics and science were easy they were somewhat divided — although nearly 50 percent agreed or strongly agreed with those statements. This broad admission of boredom and agreement that mathematics and science were easy could simply belie their other responses on enjoying the subjects, liking them, and doing well in them.

Alternatively, these latter data may suggest that students truly did enjoy the subjects, like them, felt they did well in them, and thought they were easy, even when bored by them. This would be typical of simple, unchallenging work that might seem easy, as well as bore easily, but be geared for student enjoyment and securing their involvement. With no criteria other than comparing themselves with each other, they might even feel they were performing acceptably without realizing how short they fell of the accomplishments of students in other TIMSS countries.

Students' personal attitudes relate complexly to their peers attitudes and their parents attitudes (here represented by their mothers) as the data displayed in Exhibit 4.10 shows. For example, when asked how important they felt it was to do well in mathematics and science, over half of them strongly agreed that it was important and, when asked how their mother felt about this importance, gave a similar assessment. In startling contrast, less than 15 percent strongly agreed when asked whether their friends thought mathematics and science were important.

This suggests that students reported what they felt they ought to report about their attitudes towards mathematics and its importance, attitudes communicated by parents and those in authority. At the same time, their responses strongly suggested that peer opinions on the matter were far different. On the side of honesty in reporting, however, is their reports on what their mother, themselves, and their friends felt about being good at sports and having time for fun. Here the students' attitudes echoed those of their peers even when they differed markedly from their mothers' attitudes.

The impact of peer opinion is a central fact of life in adolescence. This widespread perception of peer opinion about the importance of mathematics and science was not likely without consequences for students' own views. At worst, these students' reports of their perceptions of peer attitudes reflected their own beliefs and what they reported as their own was more what they thought parents and others would want them to believe. At best, they likely felt a conflict between the attitudes of those in authority and, perhaps, even their own reactions compared to what they believed their peers attitudes to be. As Exhibit 4.10 demonstrates, this conflict was certainly not unique to American adolescents.

In contrast, when thirteen-year-olds were asked to rate working hard, having talent, and good luck as reasons for success in mathematics and science, there were marked cross-national differences in student responses. Over half of US students indicated that hard work was more important than talent or luck. This was also the dominant belief in Germany and Japan, but not by as high a proportion. More German than US students believed talent to be more important than hard work or luck. These US students seemed to have a somewhat different view – perhaps a traditionally American view – of how to do well in mathematics and science: work hard and use whatever talent you've been given.

Students also differed in how they used their time outside of school, as Exhibit 4.11 makes clear. Certainly how students use their out-of-class time has the potential to affect how much they learn, especially when homework, studying, and attending additional classes are considered. Korean, Belgian

Exhibit 4.10. *Percentage of Students Responding 'Strongly Agree' that They, Most of Their Friends, or Their Mother Thinks Various Things Are Important.*

	Doing Well in Science			Doing Well in Mathematics		
	Friends	Self	Mother	Friends	Self	Mother
USA	11.0 (0.6)	54.3 (1.3)	60.3 (1.0)	14.5 (0.6)	61.4 (1.1)	69.6 (1.0)
Australia	7.5 (0.5)	34.9 (0.9)	33.9 (0.9)	14.7 (0.7)	50.4 (1.0)	56.0 (0.9)
Belgium (Fl)	10.4 (0.9)	37.1 (1.4)	34.3 (1.3)	23.5 (1.6)	57.0 (1.5)	59.1 (1.3)
Canada	9.2 (0.6)	43.5 (1.0)	50.3 (1.2)	19.0 (0.9)	60.2 (0.9)	71.2 (0.9)
Czech Republic	10.3 (0.6)	27.3 (1.2)	29.9 (1.1)	25.3 (1.2)	48.8 (1.6)	54.1 (1.6)
England	17.6 (1.2)	55.0 (1.3)	42.1 (1.4)	27.0 (1.5)	67.4 (1.4)	61.2 (1.4)
France	10.1 (0.6)	31.6 (1.2)	31.1 (1.2)	36.9 (1.2)	67.5 (1.3)	73.6 (1.2)
Germany	10.3 (0.7)	35.9 (1.2)	33.4 (1.2)	30.7 (1.2)	63.9 (1.2)	65.8 (1.2)
Hong Kong	15.5 (0.9)	36.4 (1.1)	24.7 (1.0)	30.3 (1.2)	56.3 (1.3)	49.8 (1.3)
Hungary	8.8 (0.6)	23.1 (1.0)	21.0 (1.0)	14.4 (0.7)	32.3 (1.1)	37.6 (1.2)
Israel	16.8 (2.0)	44.1 (2.3)	48.1 (2.4)	54.5 (2.6)	82.8 (1.5)	84.1 (1.3)
Japan	12.7 (0.6)	22.8 (0.7)		26.0 (0.9)	38.4 (0.9)	
Korea	18.6 (0.9)	34.2 (1.2)	26.9 (0.9)	34.8 (1.2)	50.3 (1.1)	56.9 (1.1)
Netherlands	20.1 (1.1)	42.9 (1.6)	36.2 (1.4)	23.4 (1.3)	48.3 (1.7)	42.7 (1.6)
New Zealand	9.9 (0.8)	43.8 (1.2)	39.2 (1.1)	16.3 (1.0)	58.1 (1.2)	57.9 (1.1)
Norway	11.5 (0.7)	35.8 (1.2)	33.9 (1.0)	20.1 (0.9)	50.7 (1.4)	50.7 (1.2)
Russian Federation	26.2 (0.9)	49.4 (0.9)	48.4 (0.9)	32.4 (1.0)	56.8 (1.1)	56.5 (1.1)
Singapore	40.8 (1.6)	63.3 (1.3)	48.3 (1.5)	46.5 (1.4)	69.4 (0.9)	59.9 (1.2)
Spain	28.2 (1.1)	62.8 (1.1)	68.0 (1.0)	32.2 (1.2)	67.2 (1.0)	75.1 (0.9)
Sweden	9.0 (0.8)	25.2 (1.1)	26.9 (1.2)	12.5 (0.8)	34.1 (1.2)	35.3 (1.2)
Switzerland	8.7 (0.6)	25.7 (0.9)	20.4 (0.8)	38.1 (1.2)	64.6 (1.2)	59.3 (1.2)
Thailand	42.8 (1.6)	62.8 (1.3)	63.0 (1.3)	44.9 (1.6)	64.1 (1.3)	67.1 (1.2)

	Be Good at Sports			Have Time to Have Fun		
	Friends	Self	Mother	Friends	Self	Mother
USA	79.6 (0.7)	76.6 (0.8)	27.5 (0.7)	54.4 (1.2)	57.0 (1.0)	42.4 (1.1)
Australia	73.5 (0.8)	70.1 (0.8)	21.8 (0.8)	32.1 (1.0)	44.6 (0.8)	37.3 (0.8)
Belgium (Fl)	67.9 (1.0)	67.4 (1.0)	21.3 (1.3)	26.6 (1.3)	38.9 (1.5)	44.8 (1.5)
Canada	78.1 (0.7)	75.1 (0.8)	22.0 (0.8)	42.6 (0.8)	47.8 (0.9)	42.6 (0.9)
Czech Republic	62.4 (1.2)	57.4 (1.3)	21.0 (1.4)	31.0 (2.0)	38.5 (1.7)	27.1 (1.1)
England	76.0 (1.0)	69.8 (1.2)	20.1 (1.2)	33.9 (1.6)	40.1 (1.7)	43.8 (1.4)
France	72.4 (0.9)	63.1 (1.1)	15.8 (0.8)	32.5 (1.1)	36.3 (1.1)	34.3 (1.1)
Germany	76.6 (1.1)	80.1 (0.9)	21.1 (1.0)	29.2 (1.2)	42.8 (1.0)	56.9 (1.1)
Hong Kong	46.1 (0.9)	45.6 (1.2)	12.5 (0.8)	22.5 (0.9)	34.7 (1.1)	11.2 (0.6)
Hungary	40.3 (1.0)	50.9 (1.0)	27.3 (0.9)	24.5 (1.0)	32.7 (1.1)	48.6 (1.1)
Israel	77.0 (1.4)	77.8 (1.5)	29.1 (1.6)	33.1 (2.3)	46.6 (2.1)	54.0 (1.8)
Japan	61.9 (0.6)	70.7 (0.6)		20.9 (0.8)	31.9 (0.7)	
Korea	38.7 (1.1)	32.3 (1.2)	17.1 (0.7)	23.8 (1.1)	34.7 (1.2)	9.2 (0.5)
Netherlands	56.8 (1.2)	69.8 (1.5)	14.7 (0.9)	15.7 (1.4)	35.2 (1.4)	49.3 (1.2)
New Zealand	74.4 (1.1)	73.6 (1.1)	31.2 (1.0)	41.4 (1.3)	50.8 (1.2)	45.9 (1.0)
Norway	74.7 (0.8)	81.7 (0.8)	18.6 (0.8)	33.4 (1.2)	40.9 (1.1)	59.1 (1.0)
Russian Federation	64.0 (1.1)	63.4 (1.3)	38.8 (1.0)	38.4 (0.9)	49.2 (0.9)	44.3 (1.2)
Singapore	42.9 (1.3)	43.7 (1.4)	9.8 (0.7)	23.4 (1.0)	36.1 (0.8)	9.3 (0.6)
Spain	77.0 (0.7)	76.1 (0.8)	34.3 (0.9)	49.0 (0.9)	58.3 (1.1)	41.7 (1.0)
Sweden	68.5 (1.0)	78.3 (0.8)	25.0 (1.0)	19.8 (0.9)	37.5 (1.0)	59.2 (1.0)
Switzerland	65.9 (1.1)	69.6 (1.2)	22.3 (0.9)	34.0 (1.1)	44.7 (1.0)	42.3 (1.1)
Thailand	39.7 (1.4)	40.1 (1.5)	40.0 (1.0)	34.4 (1.1)	46.0 (1.0)	24.2 (0.9)

(Flemish), and Singaporean students spent much more time in these out-of-class learning-oriented activities than did their US counterparts. In fact, most nations' students spent more time studying than did the US students. US eighth grade students spent more time in job-related and sports activities than did students from most other countries. However, none of these differences were very large.

Exhibit 4.11*. Eighth Grade Students' Use of Out-of-Class Time.*

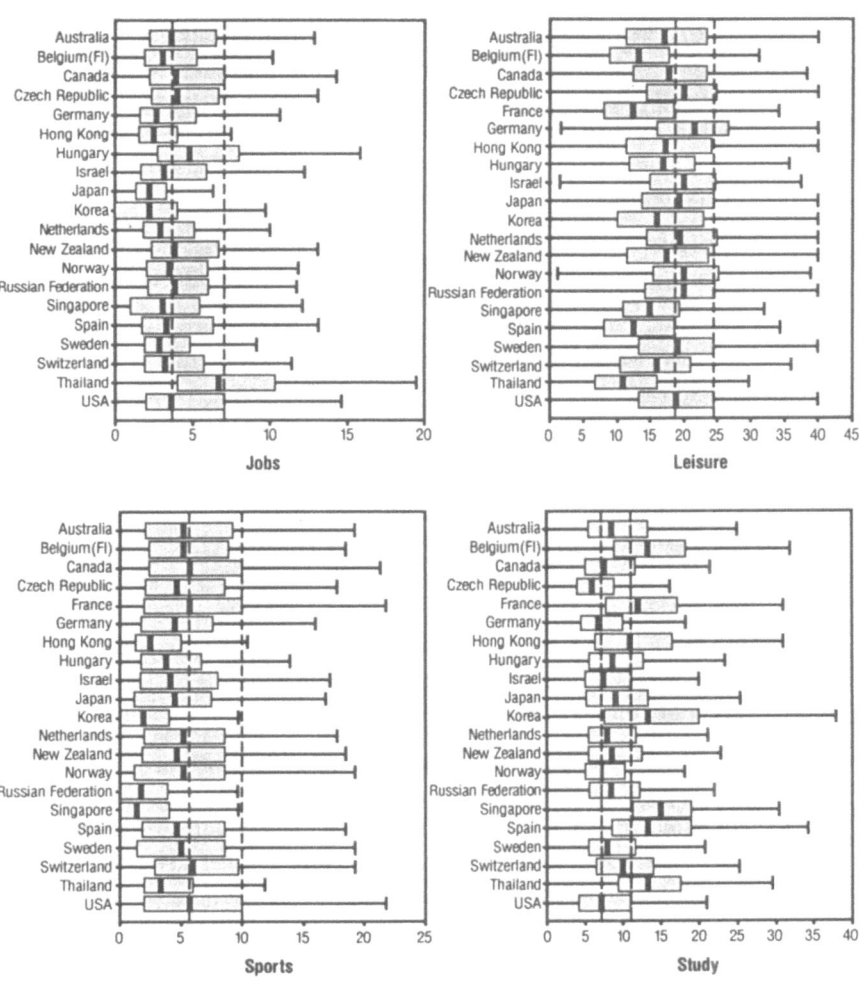

STUDENT FACTORS: A SUMMARY

We surveyed various aspects of students beliefs and attitudes to look for further factors that might affect students' utilization of learning opportunities, regardless of delivered differences in access to such educational possibilities. Our findings include:

- US students were not that different from those in other TIMSS countries in both their beliefs and behaviors.

- US students studied less, were slightly more optimistic, and more often worked at after-school jobs.

- US students believed more often that hard work was what resulted in higher achievement.

- The story behind the US' disappointing mathematics and science achievements in TIMSS does not seem to be primarily a story about US students. Although there were some differences, especially those related to using educational opportunities such as more absenteeism and less study time, US students were simply not that different from students in other TIMSS countries.

- Absenteeism and tardiness were widespread problems in US schools, and could affect student access to educational possibilities.

- In spite of much anecdotal evidence and reportage, more serious offenses such as drugs and violence by students were highly localized in a few schools but within those schools the problem was serious.

- In the US, as in other countries, peer culture seems to have continued to draw students away from valuing school and the things that go with it.

OTHER CONSEQUENCES: CONCLUDING REMARKS

This chapter has surveyed other systemic factors – those of school organization, teacher assignments, teacher beliefs, and student attitudes that might be related to the science and mathematics achievement of US students. We focused especially on seventh and eighth grade students.

US seventh and eighth grade teachers differed in whether they viewed mathematics and science in formal ways, and in whether they emphasized real-world applications of the disciplines they taught. The patterns of US teachers differed somewhat from those of their counterparts in other TIMSS countries. These differences among US teachers reflected some commonalities, but also reflected broad enough differences to suggest that US education systems were splintered in how they trained and prepared teachers as well as in the attitudes and beliefs teachers came to have as they worked within these systems. More direct questioning showed US teachers typically to have sought one or a few approaches to an instructional situation while teachers in some other countries, most notably Japan, characteristically appeared to consider a wide variety of approaches to the same situation.

US schools often used at least *de facto* specialist teachers for mathematics and science, that is, teachers with assignments devoted primarily to mathematics or science. Other countries did so as well. Much of seventh and eighth grade mathematics and science instruction was not provided by teachers with such specialized assignments. This perhaps correlates with another finding: most US schools that contained seventh and eighth grades did not contain grades higher than eighth. That is, these grades in the US (but not in many other countries, for example, Korea, Singapore, and the Netherlands) were grouped in schools as if they were the culminating grades of elementary school rather than the initiating grades of secondary school. This likely had consequences for teacher assignments, teacher training and experience, the level of expectations for students, and the level of challenge in the mathematics and science curricula US seventh and eighth graders encountered.

US principals reported their time was devoted much more often to external and internal relations than to routine administrative tasks and teaching (which contrasted sharply with many other countries). US principals seemed most to have functioned as 'managers' in an organization with responsibilities for public relations as much, or more, than responsibilities for the 'product' produced or 'service' provided by their organization.

The beliefs, attitudes, and behaviors of US students could have inhibited utilization of educational possibilities. US students reported studying less, but being more optimistic about how well they were doing in mathematics and science than students in other countries, perhaps reflecting differences in the school mathematics and school science they experienced. They appeared to be in (likely unconscious) conflict over whether to have attitudes more like their parents or more like their peers. In some cases, their reported attitudes matched that of their mothers; in other cases, that of their friends. This was especially relevant to the importance of mathematics and science in which parent and peer attitudes differed widely. This conflict may also have been the cause of the large proportions of US seventh and eighth graders who reported finding mathematics enjoyable but also boring. Such a conflict certainly would hinder student utilization of access to educational possibilities.

US students believed more than their counterparts in other countries in the efficacy of hard work in learning mathematics and science. They did this at the same time that they devoted less out-of-class time to studying and more to job-related activities. Students in most TIMSS countries devoted more out-of-class time to leisure activities than either academic or job matters.

US schools more often had to deal with student problems such as tardiness and absenteeism which also affected student utilization of educational opportunities. On the other hand, US seventh and eighth graders did not differ significantly in more serious problems of violence, substance abuse, inappropriate sexual behavior, etc. Serious problems in the US were confined to a small proportion of schools and, within those schools, to a small percentage of students.

Overall, there are differences in US schools, teachers, and students that could contribute to differences in cross-national comparative achievements in mathematics and science, although most of these seem likely to have had limited impact. Three of the more serious, in the context of the US system, seem to be a 'managed organization' rather than an achievement- or learning-centered conception of schools, the grouping of seventh and eighth grades with elementary school rather than with secondary school, and the smaller amount of study time by US students. These kinds of differences seem to be systemic and reflect characteristic features of US education systems and underlying values, rather than being solely the personal weaknesses of students, teachers, or principals. Even US students' comparatively lower out-of-school study time reflects, at least in part, the unchallenging and repetitive nature of the US curricula.

Notes–

[1]   See Stigler, et al. (in preparation). *The TIMSS Classroom Videotape Study.* Department of Education.   National Center for Education Statistics: Washington, DC.   See also: http://nces.ed.gov/timss/video/index.html.

[2]   See Schmidt, W.H. et al. (1966). *Characterizing Pedagogical Flow: An Investigation of Mathematics and Science Teaching in Six Countries.* Dordrecht: Kluwer.

Part II
Our Students' Accomplishments

We have looked at some characteristics of US education systems – what is taught in mathematics and the sciences, to whom it is taught, and how it is taught. We turn now to a description of our students' accomplishments as they emerged in the TIMSS achievement results thus far released. We (the authors) move beyond global, highly aggregated reports of achievement comparisons to examine these results in more detail. We seek to examine the results in ways that are more sensitive to the effects of mathematics and science curricula, so that if curricula really do matter to achievement, we will understand better how this works.

In Chapter 5, 'Curriculum Does Matter,' we discuss the fact that, contrary to appearances with highly aggregated, scaled test results, achievement *does* relate to curricular differences. This should seem obvious. If achievement is not related to curricula and what happens in school, why gather achievement data at all? Surely it is not for the purpose of declaring winners and losers. The US and most other countries participate in cross-national studies such as TIMSS to learn more about what is working well and what is working less well in their education systems. If achievement scores cannot be related to educational factors that can be changed (that is, cannot be related to making better educational choices), there seems little point in going through the massive efforts needed to gather cross-national educational information.

In Chapter 6, 'Access to Curriculum Matters,' we go further. The data suggest more than that achievement differences are related to curricula. The data suggest that the differences are also related to students having access to different possibilities for learning mathematics and science. This differential access seems strongly related to varying achievement. If true, this not only further shows that curriculum matters but calls into serious question policies leading to such differences in access to educational possibilities. We believe the data show that curricular differences are related to differences in access to educational possibilities and that these do matter. They matter to overall student achievement. They also affect how much students can possibly achieve even by working hard and to the best of their abilities.

Chapter 5
Curriculum Does Matter

Often achievement testing as a part of cross-national comparative studies has been treated as an end in itself. Perhaps for some countries documenting their comparative status among other countries serves sufficient political functions to justify the costs and effort necessary to gather these achievement comparisons. Perhaps this is enough for countries that are concerned primarily about within-country distributions of achievement as they relate to educational resource allocations and to other social and political factors. Most countries who take part in large scale comparative studies, however, are interested in more than politically useful status reports. A deeper justification for such studies is the light they can shed on the comparative strengths and weaknesses of education systems, light that allows a more informed pursuit of systemic change.

WHY CURRICULUM APPEARS NOT TO MATTER

Cross-national comparisons have often seemed to reveal few achievement differences that were attributable to specific aspects of curriculum, or even schooling. This appearance stems more from the tendency of such studies to look only at the 'big picture' – that is, to look at scaled global scores that combine many aspects of subject matter content. This chapter tries to demonstrate an alternate way to use cross-national achievement data. It hinges on making achievement comparisons increasingly specific. When this can be done – and can be done while still meeting reasonable statistical and psychometric criteria for significant differences – it turns out to relate to curriculum and other education system factors more directly. In that sense, presenting such analyses demonstrates that curriculum *does* matter.

How can we make comparisons more specific and, indirectly, more sensitive to curricular factors? Much achievement reporting uses scores that are scaled from combinations of many subject matter elements – for example, many different topics and sub-topics within mathematics. Many modern scaling techniques – for example, Rasch scaling that is used in some TIMSS achievement reporting – have used sets of items to estimate a score on a single underlying trait. That trait, an ability or achievement, is assumed to reside uniformly in the combination of items used to make the estimate. If items come from several areas – from geometry, algebra, and more numerical topics in mathematics – the underlying trait reflects only what is common to these topics, a kind of general mathematics or problem-solving ability. Specific differences among the contents in items combined is treated as a form of error or 'noise.' Unfortunately, it is precisely these content-specific differences among items that make achievement assessments curricularly sensitive and to label them as 'lack of fit' or 'error' is to ensure that the resulting scaled scores are unlikely to relate to specific features of curricula and educational systems.

Can we find evidence that the apparent lack of curriculum effects in cross-national studies is partly, if not largely, related to scaling across disparate subject matter contents? We believe the answer is 'yes.' We believe that we find specific content effects by identifying strong interactions of particular items with countries (that is, item scores that systematically differ far more greatly among countries than do scale scores based on several items).[1] We believe that we can also find curricular effects by examining variations of scores for sub-test areas – that is by testing two-dimensional geometry basics or earth features rather than mathematics or science.

This kind of specification overcomes the tendency of aggregation and scaling to deaden curriculum sensitivity by bringing out only the commonalities that underlie clusters of disparate items. Here clusters of items are still used, but they are less disparate in content and any underlying 'trait' estimated represents something that has more meaning in terms of the subject matter involved. This kind of analysis in which one country may perform better in some content areas but less well in others also tends to demonstrate that it is extremely difficult for one country to 'have it all,' that is, have high achievement scores in all areas of a test covering a broad range of content.

We believe that we can find curricularly sensitive achievement differences by looking at gains over a single grade's instruction. Achievement scores based on a single point in time, even for specific content sub-areas, represent cumulative achievements. Establishing relationships between curriculum and instruction measures at a given grade level and such cumulative achievement measures is complicated, if not impossible.

However, if achievement in a sub-area is estimated at two points in time, approximately one 'grade' apart, the gains in achievement from the first to the second point are more directly relatable to a single grade's intervening instruction. It is easier to get specific instructional and curricular information for a single grade and since focusing on this comparatively small 'slice' of educational time minimizes cumulative differences among countries, such gain results are more likely to be sensitive to curricular factors. Unfortunately, the TIMSS design does not allow tracking the same students from beginning to end of a single grade. However, it does permit estimating national or sub-national scores at one grade (for example, the lower of two adjacent grades containing thirteen-year-olds) and comparing them to comparable scores at the next higher grade (for example, the upper of two adjacent grades containing thirteen-year-olds).

This chapter explores the results of applying these various approaches to moving beyond achievement findings that are not sensitive to curriculum. It focuses on the status of US students' science and mathematics achievements compared to those of other TIMSS countries. It also focuses on the gains made by US students across the two pairs of adjacent grades at which assessment data were gathered (third and fourth for nine-year-olds, and seventh and eighth for thirteen-year-olds).

ASSESSING THE COMPARATIVE STATUS OF US STUDENTS

The first set of analyses uses various approaches to examine the comparative status of US students in achievement related to aspects of mathematics and science. It seeks to do so with sufficient specificity to provide curricularly sensitive achievement measures.

COMPARATIVE ACHIEVEMENT FOR SPECIFIC CONTENT AREAS

TIMSS achievement reporting thus far has been limited to global mathematics and science scale scores and to reporting the national percentages of items correct in a set of six 'reporting categories' in both subjects. These reporting categories were still so broad – as the global scale scores obviously were – as to include somewhat disparate items. Our first analyses focus on reports for more specific content areas.

Fourth grade mathematics. Exhibit 5.1 presents achievement scores for the TIMSS countries that administered the achievement test for the upper of the two grades containing nine-year-olds (fourth grade in the US). The scores are presented for 14 specific content areas within mathematics. Each country's score for an area represents the mean percent of the items in that area that were answered correctly by that country's students. The countries are listed from the highest scoring to the lowest in each content area. Each area's scores are divided by shading into three tiers – those countries that scored significantly higher than the US, those whose scores were not significantly different from that of the US, and those that scored significantly lower than the US.[2]

Our comparatively best performance was in the area of 'geometry: position and shapes' and in 'symmetry, congruence and similarity.' In the first, only Australia scored significantly higher than the US and in the second only three countries' students outperformed US students. US students also did well in

Exhibit 5.1. *Mathematics Scores for Specific Content Areas for Fourth Grade* Students Compared to the US (national percent correct in each area).***

Meaning of Whole Numbers	
Korea	88.1
Singapore	81.6
Japan	79.9
Hong Kong	76.4
Hungary	74.5
Netherlands	73.3
Czech Republic	72.9
Israel	69.8
USA •••••	69.8
Australia	68.9
Canada	68.7
International	65.9
Norway	61.4
New Zealand	59.1
England	59.0
Thailand	57.3

Whole Number Operations	
Korea	82.5
Japan	80.2
Singapore	79.2
Hong Kong	76.9
Czech Republic	74.4
Netherlands	73.2
Hungary	71.5
USA •••••	67.6
Israel	66.3
International	64.1
Australia	61.5
Canada	61.5
Norway	57.6
England	53.1
Thailand	52.7
New Zealand	50.8

Common Fractions	
Singapore	71.8
Hong Kong	65.3
Japan	62.5
Korea	61.5
Hungary	61.2
Netherlands	60.2
USA •••••	53.7
Israel	53.3
Australia	52.0
Canada	50.4
International	50.0
Czech Republic	49.7
England	49.1
Thailand	46.4
New Zealand	45.3
Norway	39.2

Decimal Fractions	
Singapore	83.2
Korea	75.0
Hong Kong	72.6
Japan	71.3
Netherlands	47.1
USA •••••	46.8
International	44.6
Australia	43.7
Thailand	43.6
Canada	41.5
Czech Republic	40.1
Hungary	36.4
Israel	34.2
England	32.0
Norway	28.9
New Zealand	26.3

Perimeter, Area, & Volume	
Singapore	75.9
Korea	73.5
Netherlands	71.0
Hong Kong	70.8
Czech Republic	68.2
Japan	67.5
Hungary	62.8
Australia	62.2
Canada	59.3
International	58.8
USA •••••	58.3
Israel	57.1
England	55.2
Norway	54.5
New Zealand	49.1
Thailand	48.2

Geometry: Position & Shapes	
Australia	71.9
England	70.6
Netherlands	70.3
Hong Kong	70.1
Canada	69.4
USA •••••	67.7
Czech Republic	67.4
Japan	66.7
Singapore	65.3
New Zealand	65.3
Korea	65.1
Hungary	62.3
International	61.7
Israel	59.1
Norway	58.3
Thailand	47.6

Symmetry, Congruence, & Similarity	
Singapore	88.9
Korea	88.8
Hong Kong	85.4
England	80.9
Australia	79.9
Japan	79.5
USA •••••	78.5
Canada	78.5
Czech Republic	77.8
Netherlands	74.1
Hungary	73.3
International	70.5
New Zealand	69.1
Israel	69.0
Thailand	66.5
Norway	56.1

Proportionality	
Singapore	63.7
Netherlands	62.7
Korea	58.1
Japan	57.7
Czech Republic	57.3
Hungary	55.5
Hong Kong	53.3
Australia	51.3
England	46.6
Israel	46.0
International	45.5
USA •••••	45.5
Canada	43.8
New Zealand	43.7
Norway	40.8
Thailand	34.0

*   See note 3 on page 12.
** International means are based on all countries participating in the TIMSS assessment at each of the indicated student grade levels. Significant difference categories were determined using standard errors and comparisons among the participating countries. Only the focal countries for this report are included in the exhibits. Therefore, each significant difference category may contain additional countries to those included in the exhibit tables.

'data representation, probability and statistics' (outscored only by Singapore, Korea, and Japan).

Our worst scores were in 'measurement units' and 'estimating quantity and size.' In both cases we were well below the international average with none or one of the 14 countries reported here ranked below us. Only these two areas required knowledge of metric system basics (the latter because the sizes estimated were in metric units and the former because it included questions about which metric units were most appropriate for certain measurements). Another area in which our performance was relatively poor was 'proportionality' in

Exhibit 5.1. *(Continued).*

Estimating Quantity & Size	
Japan	78.0
Hong Kong	70.5
Czech Republic	67.9
Hungary	67.3
Korea	67.0
Netherlands	65.9
Singapore	65.3
Australia	61.2
Norway	60.6
International	57.2
England	52.2
New Zealand	51.9
Canada	50.7
Israel	50.5
Thailand	49.8
USA •••••	47.1

Rounding & Estimating Computations	
Korea	76.8
Japan	74.8
Hong Kong	74.8
Singapore	74.3
Netherlands	68.7
Hungary	68.6
Czech Republic	68.1
USA •••••	67.2
Canada	61.6
Israel	61.2
Australia	59.4
International	55.2
New Zealand	47.9
England	47.7
Norway	46.4
Thailand	39.1

Measurement Units	
Japan	76.5
Korea	72.1
Netherlands	72.0
Czech Republic	68.2
Hong Kong	67.6
Hungary	66.3
Singapore	61.3
Australia	60.2
Norway	60.0
International	56.8
England	52.3
Canada	51.7
Israel	51.0
New Zealand	49.7
USA •••••	48.0
Thailand	43.2

Patterns, Relations, & Functions	
Korea	85.0
Hong Kong	77.8
Singapore	77.5
Japan	77.2
Hungary	71.4
Netherlands	70.4
Czech Republic	69.8
Australia	69.6
USA •••••	69.1
Canada	67.1
International	63.2
Israel	62.6
England	62.0
New Zealand	57.5
Norway	54.7
Thailand	52.4

Equations & Formulas	
Korea	85.3
Japan	84.3
Singapore	82.6
Hong Kong	76.9
Hungary	72.4
Netherlands	72.2
Czech Republic	71.0
Israel	66.8
USA •••••	65.6
International	62.0
Canada	61.4
Australia	60.0
Thailand	54.0
New Zealand	50.5
Norway	49.5
England	48.2

Data Representation Probability & Statistics	
Singapore	80.8
Korea	79.8
Japan	78.3
Hong Kong	75.8
Netherlands	75.4
USA •••••	73.0
Canada	68.0
Czech Republic	67.1
Australia	67.1
England	65.8
Israel	63.4
International	61.8
New Zealand	61.5
Hungary	60.3
Norway	59.5
Thailand	55.5

☐	Significantly Higher
☐	No Significant Difference
▨	Significantly Lower

which we had several countries above us and only two below us. In the numerical 'basics' (meaning of whole numbers, whole number operations, and common fractions) we were above the international average but they were still not the areas in which we had our relatively best achievement.

Fourth grade science. Exhibit 5.2 presents comparable data for 15 areas of science for the upper of the two grades containing most nine-year-olds (fourth grade in the US). No countries outperformed US students in the areas: 'organs and tissues', 'interactions of living things', 'human biology and health', and 'scientific processes.' Our students' average scores were higher than the

Exhibit 5.2. *Science Scores for Specific Content Areas for Fourth Grade Students Compared to the US (national percent correct in each area).*

Earth Features	
Korea	70.9
Japan	63.3
USA •••••	60.2
England	59.1
Canada	58.4
Czech Republic	57.9
Hungary	57.6
Netherlands	57.3
Norway	56.3
Australia	55.8
New Zealand	54.0
Hong Kong	53.7
International	53.3
Singapore	52.7
Israel	48.4
Thailand	42.5

Earth Processes	
Korea	66.8
Japan	61.6
USA •••••	60.8
Netherlands	58.3
Australia	55.8
England	55.1
Canada	54.8
Singapore	54.3
Czech Republic	53.8
Hungary	53.4
New Zealand	52.7
International	49.7
Hong Kong	51.5
Israel	47.8
Thailand	44.9
Norway	42.2

Earth in the Universe	
Czech Republic	75.5
Hong Kong	75.2
Hungary	74.9
Norway	74.7
Korea	73.8
Australia	72.3
Canada	70.7
England	70.5
USA •••••	70.5
Netherlands	68.7
Japan	68.7
Singapore	68.6
International	66.8
New Zealand	66.3
Thailand	60.8
Israel	58.2

Plants & Animals	
Korea	81.4
Japan	78.0
Australia	72.4
Czech Republic	72.2
Singapore	70.4
USA •••••	69.6
Netherlands	67.5
Canada	67.3
England	67.1
New Zealand	66.8
Hungary	65.9
Hong Kong	64.7
International	63.6
Norway	59.9
Thailand	56.5
Israel	55.0

Human Biology & Health	
Netherlands	71.8
Korea	70.4
Australia	70.2
USA •••••	69.3
Norway	68.4
Singapore	67.5
England	67.0
Hong Kong	67.0
Canada	66.4
Czech Republic	66.2
Japan	66.1
Hungary	65.0
New Zealand	63.5
International	63.2
Israel	62.2
Thailand	52.1

Matter	
Korea	77.3
Japan	75.4
Netherlands	67.7
Singapore	66.9
Australia	62.6
USA •••••	61.5
Canada	61.0
Czech Republic	60.7
Hong Kong	60.3
Hungary	59.1
England	58.0
International	57.8
New Zealand	57.6
Norway	56.0
Israel	52.9
Thailand	46.4

Energy & Physical Processes	
Korea	72.9
Japan	70.1
Australia	62.6
Netherlands	62.1
England	61.5
Canada	61.1
Czech Republic	61.1
USA •••••	60.7
Singapore	60.7
Hong Kong	59.0
Israel	57.9
Hungary	57.2
International	56.3
New Zealand	55.5
Norway	52.7
Thailand	43.9

Physical & Chemical Changes	
Netherlands	73.7
Korea	72.1
Japan	71.7
Hong Kong	67.6
Czech Republic	67.1
USA •••••	64.7
Singapore	64.7
Australia	63.5
Canada	61.8
England	60.6
Hungary	59.9
International	59.5
Norway	58.2
Israel	57.5
New Zealand	56.8
Thailand	46.2

international means in all but four areas: 'earth in the universe', 'life cycles and genetics', 'matter', and 'forces and motion.'

Eighth grade mathematics. Exhibit 5.3 presents data similar to that of Exhibit 5.1 but for the upper of the two grades containing the most thirteen-year-olds (eighth grade in the US) and the 20 topic areas which were included on the test at this level. Our two relatively strongest content areas at this level were 'rounding' and 'estimating computations.' These are topics that combine traditional content on rounding with a current reform emphasis on estimating computations which becomes particularly important with increased calculator use. Only four countries (Singapore, Canada, Korea, and the Czech Republic) significantly outperformed the US in the area of 'rounding.' This proved to be one of our better areas in the fourth grade tests as well (the two areas are combined into one at the fourth grade level).

Exhibit 5.2. *(Continued).*

Organs & Tissues	
Singapore	66.6
Australia	66.4
Czech Republic	65.3
USA • • • • •	65.0
Hong Kong	64.6
Korea	63.4
Japan	62.9
England	61.6
New Zealand	60.8
Canada	61.4
Hungary	61.1
Netherlands	60.8
Norway	60.5
International	58.9
Israel	56.7
Thailand	55.0

Life Processes & Functions	
Korea	75.5
Japan	75.0
Netherlands	74.0
Australia	69.9
Czech Republic	69.4
England	69.0
Singapore	68.5
USA • • • • •	68.2
Norway	67.2
Israel	66.8
Canada	66.4
Hong Kong	65.6
Hungary	64.3
International	64.3
New Zealand	62.7
Thailand	47.8

Life Cycles & Genetics	
Netherlands	76.8
Korea	75.4
Czech Republic	73.1
Australia	71.3
Japan	71.2
USA • • • • •	69.7
Canada	68.6
Norway	68.4
England	67.8
Hong Kong	67.8
Hungary	67.5
Singapore	67.3
New Zealand	66.1
International	63.9
Israel	61.7
Thailand	53.1

Interactions of Living Things	
Netherlands	83.9
USA • • • • •	82.5
Australia	82.0
Korea	82.0
Japan	78.9
Canada	77.1
Norway	76.4
England	76.1
Czech Republic	75.1
Israel	75.1
Singapore	75.0
New Zealand	74.3
Hong Kong	73.9
International	71.6
Hungary	66.0
Thailand	52.9

Forces & Motion	
Korea	68.3
Japan	62.5
Czech Republic	61.6
Netherlands	61.3
Singapore	59.5
Norway	57.6
Canada	55.5
Australia	55.4
Hungary	54.6
Hong Kong	54.5
USA • • • • •	52.6
New Zealand	51.8
International	50.9
England	48.7
Thailand	46.9
Israel	45.8

Environmental & Resource Issues	
Korea	74.0
Japan	71.6
Australia	69.1
USA • • • • •	68.0
Netherlands	65.0
Czech Republic	62.7
Hungary	62.0
Canada	60.9
England	59.8
New Zealand	58.7
International	58.2
Norway	58.2
Israel	58.0
Singapore	57.2
Hong Kong	53.8
Thailand	53.4

Scientific Processes	
Korea	65.7
USA • • • • •	63.1
Australia	59.4
Singapore	54.3
England	54.1
Japan	53.3
Netherlands	52.0
New Zealand	51.7
Hong Kong	51.4
Czech Republic	51.1
Canada	50.9
Israel	49.8
Norway	48.1
International	45.6
Hungary	40.3
Thailand	40.0

☐ Significantly Higher
☐ No Significant Difference
▩ Significantly Lower

We did most poorly in two areas within the broader category of measurement: 'measurement units' (also bad at fourth grade) and 'perimeter, area, and volume.' The other area in which we had one of our lowest scores was 'polygons and circles', a part of the geometry tested.

We scored significantly above the international mean in 'rounding.' Our averages were not significantly different from the international averages for: 'uncertainty and probability', 'data representation and analysis', 'proportionality concepts', 'patterns, relations, and functions', 'equations and formulas', 'three-dimensional geometry and transformations', 'whole numbers', 'common fractions', 'decimal fractions and percentages', 'relations of fractions', 'estimating quantity and size', 'estimating computations', and 'measurement estimation and errors.'

Exhibit 5.3. *Mathematics Scores for Specific Content Areas for Eighth Grade Students Compared to the US (national percent correct in each area).*

Whole Numbers		Common Fractions		Decimal Fractions & Percents		Relations of Fractions		Estimating Quantity & Size	
Singapore	80.4	Singapore	82.4	Singapore	77.9	Singapore	85.1	Japan	77.9
Japan	72.0	Japan	72.2	Czech Republic	72.5	Korea	77.6	Singapore	77.3
Switzerland	70.2	Korea	69.9	Hong Kong	71.1	Japan	75.3	England	72.2
Belgium (Fl)	68.6	Hong Kong	69.4	Japan	69.8	Hong Kong	74.1	Hong Kong	71.8
Korea	67.6	Belgium (Fl)	68.7	Hungary	68.7	Belgium (Fl)	71.3	Switzerland	71.8
Czech Republic	67.4	Switzerland	62.4	Korea	68.4	Switzerland	70.0	Czech Republic	71.7
Hong Kong	65.4	Czech Republic	59.6	Russian Federation	65.9	France	67.2	Netherlands	71.2
France	64.5	Netherlands	58.9	France	64.9	Canada	65.1	Belgium (Fl)	70.7
Russian Federation	62.7	France	57.8	Belgium (Fl)	64.6	Czech Republic	64.9	Korea	70.7
Spain	60.8	Hungary	57.7	Switzerland	61.9	Germany	62.6	Australia	69.3
Hungary	60.6	Israel	57.4	Canada	61.0	Netherlands	62.5	Hungary	69.3
Sweden	58.9	Canada	56.2	Sweden	59.2	Sweden	62.2	Sweden	69.3
Canada	57.9	Russian Federation	55.8	Thailand	58.0	Australia	61.0	France	69.0
Israel	59.0	Australia	55.4	Netherlands	57.8	Hungary	60.7	Germany	69.0
Thailand	56.0	Thailand	54.2	Germany	57.2	Thailand	59.6	Canada	67.8
International	*55.5*	Sweden	53.1	Israel	56.3	Norway	59.6	Norway	66.6
Germany	55.4	*International*	*52.1*	USA ••••••• 55.8		Russian Federation	59.2	New Zealand	65.3
Australia	55.0	Germany	51.7	*International*	*55.4*	USA ••••••• 58.5		Russian Federation	63.5
Netherlands	53.5	New Zealand	51.5	Norway	54.1	Israel	57.6	USA ••••••• 62.7	
Norway	53.4	Norway	51.2	Australia	53.5	*International*	*56.9*	*International*	*61.0*
USA ••••••• 52.3		USA ••••••• 49.3		Spain	50.6	New Zealand	56.8	Israel	58.4
New Zealand	48.6	England	49.2	New Zealand	49.3	England	53.8	Thailand	57.1
England	46.8	Spain	49.2	England	45.9	Spain	48.2	Spain	54.8

2-D Geometry		Polygons & Circles		3-D Geometry & Transformations		Congruence & Similarity		Proportionality Concepts	
Japan	78.9	Korea	73.0	Japan	84.3	Japan	79.7	Singapore	70.1
Singapore	78.3	Japan	70.8	Singapore	78.6	Korea	76.9	Korea	57.0
Korea	75.5	Singapore	70.6	Czech Republic	77.7	Singapore	76.7	Japan	56.9
Hong Kong	73.5	Hong Kong	65.4	France	77.7	Hong Kong	71.0	Hong Kong	50.1
Russian Federation	63.4	Belgium (Fl)	64.1	Hong Kong	77.6	France	69.7	Netherlands	41.9
Netherlands	62.9	France	61.9	Switzerland	75.9	Czech Republic	66.1	Canada	40.3
Thailand	62.4	Israel	59.7	Belgium (Fl)	73.7	Thailand	62.7	Sweden	39.2
Czech Republic	62.1	Czech Republic	57.4	Korea	73.3	Russian Federation	62.3	Belgium (Fl)	39.0
Hungary	61.6	Russian Federation	57.2	Hungary	71.0	Belgium (Fl)	57.5	Australia	37.8
Australia	58.8	Thailand	53.5	Netherlands	70.6	Israel	54.8	France	37.3
Belgium (Fl)	58.8	Hungary	51.6	Russian Federation	70.2	Hungary	53.4	Switzerland	36.8
Switzerland	57.9	Canada	50.5	Canada	70.1	Canada	52.8	Czech Republic	37.3
France	56.1	Australia	50.4	New Zealand	67.3	*International*	*52.6*	Thailand	36.5
New Zealand	55.8	*International*	*48.9*	England	66.3	Switzerland	51.2	England	34.8
England	54.9	Switzerland	48.6	Australia	65.5	Australia	50.9	Hungary	34.6
Canada	54.7	England	47.0	Israel	64.8	Netherlands	51.0	Russian Federation	33.5
International	*53.8*	Germany	44.3	Thailand	64.5	Norway	46.9	*International*	*33.4*
Israel	52.5	Netherlands	44.0	Germany	63.7	New Zealand	45.9	New Zealand	33.4
Norway	52.1	New Zealand	43.2	*International*	*33.4*	Sweden	45.9	USA ••••••• 31.5	
Spain	49.5	Sweden	42.8	Norway	58.6	England	45.6	Israel	31.4
Germany	49.3	Norway	42.4	USA ••••••• 58.4		Germany	45.1	Norway	30.9
USA ••••••• 47.9		Spain	39.3	Spain	58.2	USA ••••••• 44.1		Spain	27.4
Sweden	46.1	USA ••••••• 39.1		Sweden	55.9	Spain	43.6	Germany	26.5

Our average scores were significantly below the international averages for 'measurement units', 'perimeter, area and volume', 'two-dimensional geometry', 'polygons and circles', 'congruence and similarity', and 'proportionality problems.'

Exhibit 5.3. (Continued).

Rounding		Estimating Computations		Measurement Units		Perimeter, Area, & Volume		Measurement Estimation & Errors	
Czech Republic	86.9	Singapore	75.8	Singapore	77.9	Singapore	84.5	Czech Republic	82.9
Singapore	86.6	Czech Republic	69.8	Japan	74.8	Japan	72.6	Singapore	81.0
Canada	84.7	Japan	69.0	Czech Republic	73.2	Hong Kong	63.9	Sweden	80.8
Korea	83.8	Korea	66.7	Belgium (Fl)	71.8	Korea	61.8	Korea	79.5
Belgium (Fl)	81.0	Belgium (Fl)	64.1	Switzerland	71.6	Czech Republic	48.9	Belgium (Fl)	78.9
Hungary	79.8	Switzerland	62.5	Sweden	69.5	Switzerland	46.6	Hungary	78.2
Japan	79.6	Canada	61.3	France	69.0	Belgium (Fl)	45.6	Switzerland	78.2
USA •••••••• 79.2		Sweden	60.0	Korea	67.9	Russian Federation	44.6	Netherlands	76.1
Netherlands	78.5	Australia	59.1	Hungary	67.5	France	44.1	France	75.2
Sweden	78.5	Netherlands	58.7	Netherlands	66.8	Netherlands	44.0	Australia	74.5
Australia	77.5	Hong Kong	57.8	Norway	66.4	Hungary	42.9	Norway	74.5
Switzerland	77.4	New Zealand	57.6	Russian Federation	65.9	Australia	42.1	Germany	72.9
Thailand	75.4	Hungary	57.3	Hong Kong	65.2	Canada	40.4	Hong Kong	72.7
France	74.7	Norway	57.0	Australia	65.0	Thailand	40.0	Japan	72.5
New Zealand	74.4	USA •••••••• 55.7		Germany	63.5	Sweden	39.3	England	71.8
Hong Kong	74.2	England	54.4	Canada	63.0	*International*	*39.0*	New Zealand	70.6
Russian Federation	73.5	Germany	52.0	Israel	63.0	England	37.1	Canada	69.7
England	74.2	Israel	51.9	Spain	62.4	Israel	36.9	Russian Federation	69.6
Norway	72.3	France	50.8	Thailand	62.3	Norway	36.4	Thailand	66.1
International	*70.7*	*International*	*50.6*	*International*	*61.3*	New Zealand	35.0	*International*	*64.7*
Germany	69.9	Russian Federation	50.0	New Zealand	61.3	Germany	32.9	Israel	63.3
Israel	68.8	Thailand	49.5	England	60.5	Spain	29.9	USA •••••••• 62.5	
Spain	62.2	Spain	44.1	USA •••••••• 53.0		USA •••••••• 27.8		Spain	55.3

Proportionality Problems		Patterns, Relations & Functions		Equations & Formulas		Data Representation & Analysis		Uncertainty & Probability	
Singapore	76.7	Japan	70.0	Singapore	79.8	Japan	84.3	Singapore	78.5
Japan	67.8	Korea	68.4	Japan	72.3	Korea	81.9	Korea	77.8
Korea	65.1	Singapore	67.2	Hong Kong	70.5	France	81.4	Belgium (Fl)	77.7
Hong Kong	64.0	Hong Kong	60.8	Korea	70.0	Switzerland	81.3	Japan	75.0
Czech Republic	61.3	Czech Republic	60.2	Czech Republic	64.3	Singapore	81.0	Hong Kong	73.0
Belgium (Fl)	59.4	England	60.1	Belgium (Fl)	62.1	Netherlands	80.6	Netherlands	72.4
Switzerland	57.3	Switzerland	59.5	Russian Federation	61.4	Czech Republic	79.6	Switzerland	72.4
Russian Federation	56.3	Belgium (Fl)	59.2	Hungary	60.9	Sweden	79.5	Canada	70.0
Thailand	56.0	Hungary	59.1	Israel	58.3	Belgium (Fl)	79.5	Sweden	68.7
Hungary	55.9	Canada	58.2	Thailand	54.0	Hong Kong	76.4	Australia	67.5
Netherlands	55.9	Australia	57.6	France	53.6	Germany	76.0	France	66.5
France	55.7	Netherlands	56.9	Australia	52.8	Norway	75.5	Hungary	66.2
Australia	52.2	New Zealand	55.7	Canada	52.4	Australia	75.2	New Zealand	65.9
Canada	51.6	France	55.0	Spain	52.2	Canada	75.0	England	65.3
International	*50.1*	Thailand	53.4	Switzerland	51.9	England	74.8	Norway	64.7
Sweden	49.2	Russian Federation	53.2	Netherlands	51.7	Thailand	74.5	Czech Republic	63.4
Germany	49.0	Israel	53.2	*International*	*49.4*	New Zealand	74.1	USA •••••••• 63.1	
England	47.2	*International*	*51.4*	USA •••••••• 49.4		Hungary	73.7	Germany	62.4
New Zealand	47.1	USA •••••••• 51.1		New Zealand	46.6	Israel	73.2	Israel	62.3
Norway	47.1	Spain	50.3	Germany	46.4	USA •••••••• 72.6		*International*	*59.6*
Israel	46.5	Norway	48.7	Norway	44.9	Russian Federation	72.0	Spain	58.0
USA •••••••• 45.2		Germany	47.6	England	44.1	*International*	*70.9*	Russian Federation	56.9
Spain	43.6	Sweden	47.2	Sweden	43.7	Spain	70.4	Thailand	56.5

☐	Significantly Higher
☐	No Significant Difference
▧	Significantly Lower

To summarize, the US performance was average for algebra, arithmetic, fractions, estimation, and data analysis topics. We were below average for two measurement topics, all but one geometry topic, and for 'proportionality problems.' Despite this overall poor showing, the variation in the rank of countries across these more uniform content areas indicates that curricular differences, in terms of specific contents covered and emphasized, really matter.

There were some comparisons possible between the fourth and eighth grade performances. In 'whole numbers', 'common fractions', and 'data representation and probability' we went from above average to average. We were below the international average at both grade levels for 'measurement units', and in the measurement topic of 'perimeter, area, and volume' we went from average to below average. In the geometry areas of 'congruence and similarity' and 'two-dimensional geometry' and in the algebra area of 'patterns, relations, and functions', we went from above the international average to below. In decimals we were average at both grades. In not one area did our relative performance improve from fourth to eighth grades.

Overall, we can say that there were some areas in which we were consistently average (decimals and proportionality). There were also some in which we were consistently bad ('measurement units'). There were many areas in which our standing fell, from above average to average, from average to below average, and even from above average to below average (topics in geometry and algebra). There was no area in which we were consistently good. There was also no area in which our relative standing significantly improved.

Eighth grade science. Exhibit 5.4 presents science data for the upper of the two grades with the most thirteen-year-olds. This is similar to the fourth grade science data in Exhibit 5.2. Data are presented for 17 content areas. Since our overall science score's standing was high at fourth grade and around average at eighth, we might expect another set of mostly 'falling' scores when we compare the two quasi-longitudinal cross-section scores.

In 'life cycles and genetics' we were not significantly outperformed by any country at the eighth grade level. In 'earth processes', 'earth in the universe', 'human biology and health', 'scientific processes', 'structure of matter', and 'chemical changes' we were significantly outperformed by only one or two countries. We were at our relative worst in 'physical changes' and 'forces and motion' but we did not perform significantly below the international average in any of the topic areas.

Generally, we did the best at both fourth and eighth grade in earth science, especially 'earth processes.' Although it was not strictly a 'content' area, 'scientific processes' consisted of items on which we did well at both levels. We also did well in life science at both levels although the specific content areas differed – 'organs and tissues' and the 'interactions of living things' at fourth grade, but 'life cycles and genetics' at eighth grade.

We did our worst at both levels in the physics topics – 'forces and motion', 'physical and chemical changes', and 'matter' at fourth grade, and 'physical changes' and 'forces and motion' at eighth grade. In general, physics and chemistry topics for eighth grade showed considerable variability, especially since they contained topics in which we did both our best and our worst.

Eighth grade science achievement is somewhat difficult to interpret. By eighth grade, the typical science curriculum pattern in US educational systems is to have students take a course oriented to a specific broad area of science (earth science, life science, and physical science) or, less frequently, a mixed course in 'general science.' As a result, not all students were taking or had recently studied all areas. Thus, it is hard to know whether our national averages came from general scores at the average level from students regardless of special-ization in their current course or whether they came from high scores for those whose current course covered the specific area and lower scores from those whose courses did not. As a result, this aggregate, national average score for content areas was difficult to interpret.

General conclusions about topic areas. A major conclusion to draw from the data presented above is that the fourth and eighth grade performances var-ied depending on the specific content area. However, just as importantly we might conclude that 'a test is not a test', that is, that not all tests or parts of tests are of equal relevance to every country. The US students tested and the educational systems that provided science and mathematics instruction for them had both strengths and weaknesses, as did their counterparts in other countries. However, as we changed attention among topic areas, the order of countries varied in terms of which performed better or worse.

We can use statistical techniques such as median polishing to identify topic areas in which US students exceeded expectations by using their general per-formance levels across all topic areas and considering how well most countries did in each area (taken as an indicator of whether the test was easier or more difficult for each topic area). At fourth grade, the US was below what might have been expected from the general US performance on two mathematics topics ('measurement units' and 'estimating quantity and size,' both topics that depended on knowledge of the metric system). We exceeded our typical per-formance on the test at this grade in three topic areas ('earth processes', 'organs and tissues', and 'interactions of living things'). At eighth grade there was only one area in either mathematics or science on which we did better or worse than our general performance levels would suggest (the mathematical topic of 'rounding').

Similar results would be attained for other countries. This is strong evidence that when we move from the broad to the more specific content area scores, there are definite country by topic area interactions. That is, there are specific areas for each country in which they performed better or worse than might be

Exhibit 5.4. *Science Scores for Specific Content Areas for Eighth Grade Students Compared to the US Mean (national percent correct in each area).*

Earth Features		Earth Processes		Earth in the Universe		Diversity & Structure of Living Things		Life Processes & Functions	
Czech Republic	67.1	England	66.7	Sweden	70.7	Japan	75.9	Singapore	72.3
Hungary	67.1	Singapore	66.7	Norway	70.1	Singapore	74.2	Japan	70.4
Korea	65.3	Belgium (Fl)	65.4	Netherlands	67.5	Hong Kong	73.4	Korea	68.2
Singapore	64.6	Netherlands	62.5	Czech Republic	66.9	Korea	72.5	Czech Republic	63.7
Belgium (Fl)	61.1	Norway	62.3	Japan	66.7	Czech Republic	72.3	Thailand	63.0
Sweden	60.5	USA ••••••••	60.9	Thailand	66.0	Netherlands	71.6	Belgium (Fl)	62.9
Thailand	59.7	Russian Federation	60.8	Singapore	64.9	Thailand	68.8	Netherlands	60.7
Russian Federation	59.2	Canada	60.6	Switzerland	64.8	Sweden	67.7	Hungary	59.9
Norway	59.2	Sweden	60.0	Korea	64.4	Hungary	67.1	Russian Federation	59.4
Japan	59.0	Korea	59.0	USA ••••••••	63.3	Australia	66.7	England	59.3
England	58.7	Japan	58.5	Germany	62.8	Germany	65.7	USA ••••••••	58.0
Switzerland	58.2	New Zealand	57.3	Spain	62.8	Russian Federation	65.7	Australia	57.4
Netherlands	58.0	Czech Republic	57.2	Australia	62.7	England	64.5	Canada	57.2
Australia	57.5	Australia	56.7	Belgium (Fl)	61.8	USA ••••••••	63.9	Israel	57.1
Germany	57.2	Thailand	56.0	New Zealand	61.7	Canada	63.3	Hong Kong	56.3
USA ••••••••	57.1	Switzerland	56.0	Canada	61.2	Belgium (Fl)	62.9	New Zealand	56.1
Canada	56.6	Spain	55.5	England	59.0	New Zealand	62.3	Spain	55.9
Israel	56.1	France	55.0	Hungary	58.5	Spain	61.3	Sweden	54.9
Spain	55.9	Israel	54.7	*International*	58.2	*International*	61.3	*International*	54.5
New Zealand	55.9	Germany	53.9	Hong Kong	58.0	Switzerland	61.0	Switzerland	54.2
International	55.4	*International*	53.5	Israel	56.2	Israel	60.4	Norway	54.2
France	55.1	Hong Kong	52.3	Russian Federation	56.9	Norway	59.2	France	53.8
Hong Kong	54.1	Hungary	51.3	France	53.2	France	58.4		

Structure of Matter		Energy & Physical Processes		Physical Changes		Chemical Changes		Forces & Motion	
Russian Federation	56.4	Singapore	71.2	Japan	66.7	Singapore	73.5	Czech Republic	78.1
Hungary	53.9	Japan	68.8	Singapore	63.1	Korea	65.9	Japan	73.6
Czech Republic	54.2	Korea	67.2	Czech Republic	62.9	Hungary	65.4	Singapore	73.4
Singapore	53.2	England	64.1	France	61.9	Czech Republic	64.2	Netherlands	71.2
Spain	51.0	Netherlands	63.8	Israel	61.4	England	63.7	Korea	69.6
USA ••••••••	48.2	Czech Republic	62.3	Sweden	61.3	Japan	63.5	Hong Kong	69.4
Israel	44.7	Belgium (Fl)	62.2	Netherlands	61.1	Australia	61.9	Switzerland	68.8
Sweden	42.7	Australia	61.1	Hungary	61.0	Russian Federation	61.8	Hungary	68.7
Korea	42.1	Hungary	60.2	Norway	60.0	USA ••••••••	61.4	England	67.6
Japan	41.1	New Zealand	59.9	England	58.9	Germany	60.7	Sweden	67.5
Hong Kong	40.0	Canada	59.9	Belgium (Fl)	58.7	Israel	60.6	Australia	67.2
International	39.6	Hong Kong	59.5	Canada	58.5	Canada	59.6	Norway	67.1
Australia	38.7	Israel	59.5	Korea	57.3	New Zealand	59.4	Canada	66.3
England	38.1	Germany	59.3	Australia	57.1	Hong Kong	58.2	Germany	66.9
Canada	37.1	Switzerland	58.7	Russian Federation	57.0	Spain	58.1	New Zealand	65.6
Germany	35.2	Russian Federation	58.4	Switzerland	54.0	Netherlands	57.6	Belgium (Fl)	65.2
New Zealand	34.5	USA ••••••••	57.1	Germany	53.6	Sweden	57.1	France	63.8
Norway	31.9	Norway	56.6	Thailand	53.2	*International*	56.5	Spain	62.6
Thailand	30.6	Sweden	56.6	Spain	53.1	Belgium (Fl)	55.8	Russian Federation	62.1
Netherlands	30.0	*International*	56.4	*International*	52.8	Switzerland	55.4	USA ••••••••	61.3
France	29.4	Thailand	55.3	New Zealand	52.8	Norway	53.4	*International*	61.2
Switzerland	28.6	Spain	55.2	USA ••••••••	49.1	France	52.5	Israel	54.9
Belgium (Fl)	27.4	France	54.6	Hong Kong	48.8	Thailand	50.9	Thailand	56.4

expected from their general performance. A test which happened to focus more heavily on one of these areas would result in better or worse scores for the country depending on the areas emphasized. This variability by topics also suggests that there are things to be explained about country achievements by

Exhibit 5.4. (Continued).

Life Cycles & Genetics		Interactions of Living Things		Human Biology & Health		Properties & Classification of Matter	
Czech Republic	81.8	Korea	68.8	Singapore	74.0	Japan	66.6
USA ••••••••	81.2	Singapore	65.6	Czech Republic	71.6	Singapore	66.6
England	80.4	Japan	65.0	Netherlands	71.2	Korea	65.4
Netherlands	79.3	Thailand	64.1	Japan	69.4	Czech Republic	60.4
Israel	79.0	Hungary	63.1	Belgium (Fl)	69.3	Netherlands	58.3
Belgium (Fl)	78.9	Australia	62.0	Hungary	68.7	Belgium (Fl)	57.9
Canada	78.4	England	61.5	England	68.3	Hungary	57.3
Sweden	78.2	Norway	59.9	Israel	67.6	Sweden	57.2
Norway	78.1	Canada	58.2	Germany	67.6	Hong Kong	56.4
Korea	75.9	Czech Republic	57.6	USA ••••••••	66.8	Canada	55.7
France	75.7	New Zealand	55.9	Thailand	66.5	Norway	54.8
Russian Federation	75.5	Netherlands	55.8	Australia	66.5	Australia	54.7
Germany	75.5	USA ••••••••	54.3	Korea	65.5	England	54.7
Switzerland	75.1	Germany	52.9	Canada	65.3	Germany	54.9
Japan	74.8	Belgium (Fl)	52.9	Sweden	63.9	Russian Federation	54.3
New Zealand	73.9	Spain	52.9	New Zealand	63.7	New Zealand	53.5
Hungary	73.7	Russian Federation	52.3	Spain	63.2	Israel	53.5
Australia	73.0	Sweden	52.1	Russian Federation	63.1	France	52.3
Spain	71.1	*International*	*51.0*	Norway	62.6	*International*	*51.6*
International	*70.2*	Israel	50.2	Switzerland	61.4	Switzerland	51.1
Singapore	68.6	Switzerland	49.8	*International*	*61.2*	USA ••••••••	49.9
Thailand	67.3	Hong Kong	46.6	Hong Kong	59.9	Spain	49.6
Hong Kong	65.3	France	44.8	France	55.7	Thailand	46.7

Science, Technology, & Society		Environmental & Resource Issues		Scientific Processes	
Korea	74.3	Singapore	73.1	Singapore	74.6
Hungary	73.3	Thailand	70.1	Korea	64.2
Netherlands	67.4	England	67.2	Netherlands	65.2
Sweden	66.9	Australia	64.7	France	62.2
Singapore	66.4	Netherlands	65.1	Czech Republic	60.7
Belgium (Fl)	65.2	Korea	62.7	Australia	60.6
New Zealand	60.4	Japan	61.7	Japan	60.1
Japan	60.1	Canada	61.5	England	60.0
Thailand	59.2	New Zealand	61.4	Canada	59.2
England	57.2	USA ••••••••	59.9	USA ••••••••	58.9
Hong Kong	54.8	Belgium (Fl)	59.7	Hungary	57.3
Canada	52.5	Czech Republic	59.1	Hong Kong	57.2
Switzerland	52.5	Spain	58.6	Belgium (Fl)	56.9
Norway	52.1	Norway	58.2	New Zealand	56.7
Czech Republic	51.9	Hong Kong	53.3	Thailand	55.3
Australia	51.5	*International*	*53.0*	Germany	53.4
Israel	51.3	Sweden	51.8	Israel	51.8
International	*47.9*	Israel	51.6	Switzerland	53.3
USA ••••••••	47.5	Switzerland	50.3	*International*	*53.1*
Germany	47.3	Germany	50.1	Sweden	52.4
Spain	44.1	Hungary	49.4	Norway	52.1
France	42.9	France	48.5	Russian Federation	51.2
Russian Federation	40.4	Russian Federation	46.8	Spain	49.6

☐ Significantly Higher
☐ No Significant Difference
☑ Significantly Lower

looking at their curricula in mathematics and science – by looking at what they teach, what they emphasize, and what they omit.

Eighth grade mathematics – algebra vs. non-algebra students

Data on the percent of items correct in specific content areas for US algebra students compared to other countries are now presented. In interpreting these results a caveat must be added. US algebra classes contain only approximately 20 percent of US eighth-graders and because of tracking practices these are typically the more advanced students. Therefore, such comparisons with other countries which include essentially all students are inherently biased in favor of the US. It would only be a 'fair' comparison if other countries were also to select their 'best' students. Such a comparison is explored in the next chapter. These results are examined here for their relevance to curriculum issues. More countries outperformed the US algebra students in areas of geometry and measurement (with the exception of 'three dimensional geometry and transformations') than in other areas. The US finished in the top tier of countries in six areas ('estimation of quantity and size', 'rounding', 'data representation and analysis', 'statistics and probability') and the more directly algebraic content areas of 'patterns, relations and functions' and 'equations and formulas.' The US ranked first for 'rounding.' Thus, US eighth grade algebra classes compared relatively favorably with eighth grade classes from other countries except in geometry and measurement. This may well reflect the traditional US approach of single area courses ('algebra', 'geometry', etc.) as compared to the more internationally common practice of mathematics courses that are a mixture of many contents including algebra, geometry, and other topics.

The performance of US non-algebra classes was vastly lower comparatively than the performance of US eighth grade algebra classes. Given practices of tracking, student prior achievement and ability contributed to these comparative differences as well as did curricular differences (that is, students who had performed more strongly in past grades were more often in algebra classes).

COMPARATIVE ACHIEVEMENT FOR SPECIFIC ITEMS

The movement from a single score (mathematics or science) to six broad reporting categories (algebra, etc.) to around 20 specific content areas ('congruence and similarity', etc.) produces a more varied picture of TIMSS's countries' performance and, more specifically, of US performance. What would happen if we went further and examined performance at the level of individual items?[3]

'Good Performance and Poor Performance' Items. Median polishing could again be used to identify unusual scores by comparing general (median) performance on each item across all TIMSS countries and each country's general (median) performance across all items. This could be done for both mathematics and science and for both nine-year-olds and thirteen-year-olds. The result would be a mass of data, but we could focus on only those items that

indicated a large 'country-by-item' interaction component – the items for which several countries had average achievements which were either markedly higher or lower than what would be expected from those countries' typical performance (across items) or for those items (across countries). Items for which there were such large interactions could then be examined to see if they fell into specific content areas.

These results were reported more fully elsewhere but a few findings can be summarized here.[3] In fourth grade mathematics, there were nine items on which US students did less well than would have been expected and 12 on which they did better. Of the nine, five were in a single topic area – measurement units – which is consistent with the results for topic areas discussed above. Three of the other four items were in 'estimating quantity and size', another topic area already identified as one in which US students did poorly. Essentially, most of the items in which we did less well than would have been expected from our general performance fell into one of two topic areas within the broader category of measurement.

Our strengths at this level appear to be in 'data representation, probability and statistics.' We exceeded expectations for four of the 13 test items in this content area. Overall, there were 12 items on which our performance was better than might be expected. A third of these were in this single area of 'data representation, probability, and statistics.' The other area in which there was a concentration of these items was 'rounding and estimating computations' in which we did better than expected on three of the four items.

The other five items on which we did better came from common fractions, two-dimensional geometry, and functions. In no case, however, were these 'good performance' items a significant proportion of the items in their respective content areas (as they did for the two categories discussed in more detail above).

In fourth grade science there were four items on which US students performed less well than would have been expected. Two were from the content area of 'life cycles and genetics' (but there were 14 items in this content area so, although these two represented half of our 'poor performance' items, they were far from dominating this area). The other two items were from two different content areas.

There were 10 items on which we did better than we might have been expected to do. Three of these 10 were 'scientific processes' items and two were from the area of 'environmental and resources issues.' The rest were distributed one item to a content area.

Overall, we found that item interactions for the US tended to concentrate in a few content areas, especially for mathematics but less so for science. Such concentrations seem likely to have been related to curriculum coverage or to those curriculum areas to which more students had access to learning opportunities.

Extended Response Items. Among the test items, most of which were either multiple choice or open-ended items requiring short answers (single numbers, phrases, etc.), there were some open-ended items requiring extended responses, answers that could take most of a page and require diagrams, paragraphs, complete computations, and so on. These extended response items were scored both for correctness (using a partial credit model) and for the specific type of correct or incorrect response given. Thus, these items not only contributed to overall and content area scores, but they also provided more detailed information on how students in each country dealt with each of them, whether they did so successfully or not.

These extended response items supported the same point as the item level information above. Country performance averages varied even by individual items and these differences were likely related to curriculum. This seems to have been especially true for these extended response items.

Examples are provided by two science items. The first asked when water in a tube was heated why a balloon attached to the tube increased in size. The most appropriate response, according to the international scoring committee, was that the pressure of the water vapor increased when the tube was heated. Only six percent of US students gave this response. This was true in most TIMSS countries (from around three to 15 percent giving this response) with some notable exceptions (Singapore, Korea, Bulgaria, and France, with 34, 32, 49, and 20 percent, respectively). The responses for some countries represented notable deviations from their typical achievement patterns. Kuwait, Portugal, and Iran (the former with 26 percent correct and the latter two with 19 percent correct each) had much higher than typical scores on this item.[4]

A second correct response was that water evaporates. For the US, 34 percent of the students gave this response. It was the most typical correct response for most other countries as well (Bulgaria, Kuwait, and Singapore being exceptions). For the US, the most commonly chosen incorrect response (other than 'other') was that hot air rises. This misconception was not common for other countries where the predominant incorrect response was some other explanation such as Spain's most typical incorrect response that air molecules expand when heated. Substantive differences such as these are certainly tied to curriculum. This type of item makes ties to curriculum much more obvious.

A second item asked students to explain why snow can exist on the tops of mountains while it had melted at lower levels. For the US, 50 percent of the students mentioned that it is colder on the mountain top. This is overwhelmingly the most common correct response across countries. The misconception behind incorrect responses typically reflected greater differences among the countries. Most US students who answered incorrectly (again excluding 'other' incorrect responses) indicated that the sun melted the snow. In Korea, the most common incorrect answer was that where was sunshine lower down.

In Norway, the most prevalent incorrect response was that the mountain was very high.

The data indicate that the US' performance was far from uniform or monolithic. To speak of a US performance in mathematics or science in general is to gloss over distinctions that are important to understanding how performance relates to curricular and other systemic educational factors.

The nature of these differences indicates that content areas and individual items vary in their difficulty and that the same group of students can perform well in one area and poorly in another. It is difficult to argue in the face of such variation that we should rely on generic, global scores. US fourth grade students ranged from among the TIMSS countries' best in some areas of geometry and data representation to among the worst in measurement units and estimating quantity and size. In the face of such variation, what would a fourth grade 'mathematics' score represent?

This point is made even more strongly when we consider responses to the TIMSS performance assessment items (a set of tasks in which students had to perform and physically work with problem situations rather than respond only with thought and with paper and pencil). US fourth grade science students were among the best in the world on paper and pencil tests, but those same students performed at, or near, the bottom for many of the performance tasks, even given the greatly reduced set of countries involved in these assessments. However, in all but two tasks we were not significantly different from the international mean given the high variability of most responses to these complex tasks.

A similar pattern emerged at the eighth grade level. In general, at both fourth and eighth grades, US students' performances varied greatly among the different tasks. Although our performance was rarely better than the international mean, we did do better on some tasks compared to others. Again such variance suggests the relevance of curriculum for tasks (and aggregated scores) not so generalized as to be de-contextualized and to lose their inherent links to curricular differences.

ASSESSING THE COMPARATIVE STATUS OF US STUDENTS: A SUMMARY

We have presented a selection of data about how achievement indications vary when we move from more globally scaled and aggregated scores to scores on specific content areas, items, or for categories of students such as those revealed by responses to extended response open-ended items and by performance assessment items. Among the points we sought to make were these:

Exhibit 5.5. Eighth Grade Results for an Extended Response Science Item About Air Pressure.

	Correct			Incorrect	
	10 Mentions explicitly that expansion is due to increased pressure of air/gas/water vapor when tube is heated	11 States that the water evaporates	19 Other correct	70 Mentions that hot air or gas always rises	71 Mentions that air particles or molecules expand when heated
United States	**6.4**	**33.7**	**3.1**	**5.4**	**3.9**
International	*10.4*	*41.4*	*7.3*	*6.1*	*2.6*
Australia	8.6	15.8	27.7	11.0	1.6
Austria	12.5	51.1		3.2	3.2
Belgium (Fl)	10.1	47.9	2.7	11.3	4.1
Belgium (Fr)			53.4		
Bulgaria	49.3	17.0	2.0	4.8	1.4
Canada	5.1	22.4	21.8	12.4	4.0
Colombia	1.7	46.8	1.3	0.7	0.3
Cyprus	3.2	50.7	3.7	3.3	0.5
Czech Republic	4.2	59.6	6.1	4.4	0.5
Denmark	3.5	36.8	14.5	4.1	0.7
England	3.7	50.5	15.0	5.3	2.2
France	20.3	45.2	2.9	2.1	0.4
Germany	8.5	43.9	2.6	9.0	4.4
Greece	10.5	45.0	8.2	2.2	2.0
Hong Kong	0.5	38.6	1.4	19.2	2.8
Hungary	9.9	34.3	13.2	5.0	1.7
Iceland	8.0	41.5	8.0	4.7	4.2
Iran, Islamic Republic	18.8	40.9	8.7	0.3	0.4
Ireland	4.6	39.2	14.9	9.5	1.0
Israel	15.7	57.5	3.0	8.0	1.8
Japan	1.2	56.2	9.8	1.4	5.6
Korea	31.9	35.1	8.2	3.4	5.0
Kuwait	26.1	20.3	4.6	3.9	2.7
Latvia (LSS)	13.2	40.6	2.1	9.1	3.1
Lithuania	1.4	57.9	1.7	2.0	1.1
Netherlands	3.3	47.1	0.6	6.0	4.7
New Zealand	4.7	38.3	5.0	14.2	8.5
Norway	1.5	66.1	1.8	2.2	0.4
Philippines	7.7	16.0	4.5	7.0	3.4
Portugal	19.2	27.1	5.1	1.3	1.1
Romania	10.4	41.5	2.0	4.2	1.9
Russian Federation	4.2	47.5	4.0	3.9	1.2
Scotland	3.1	39.7	4.5	8.2	3.2
Singapore	33.5	24.0	0.3	24.7	3.6
Slovak Republic	8.3	60.2	3.2	8.0	1.5
Slovenia	14.8	41.5	8.4	3.8	2.0
South Africa	5.3	6.2	5.4	3.1	4.6
Spain	7.8	56.6	2.2	2.3	3.6
Sweden	2.1	65.3	1.6	3.7	2.6
Switzerland	5.0	52.4	3.2	6.0	3.7
Thailand	18.5	38.2	7.7	4.4	1.2

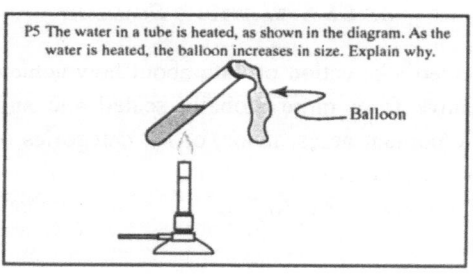

P5 The water in a tube is heated, as shown in the diagram. As the water is heated, the balloon increases in size. Explain why.

Balloon

Exhibit 5.5. *(Continued).*

Incorrect		Nonresponce		
76 Merely repeats the information in the stem	79 Other incorrect	90 Crossed out/erased, illegible, or impossible to interpret	99 Blank	
1.3	41.0	0.7	4.5	United States
2.8	21.0	0.9	9.6	International
0.8	27.8	1.0	5.7	Australia
2.0	22.5	1.8	3.5	Austria
2.0	14.8	0.3	6.9	Belgium (Fl)
	30.2		16.4	Belgium (Fr)
0.6	3.8	0.4	20.7	Bulgaria
0.2	30.3	0.1	3.6	Canada
10.1	26.5		12.5	Colombia
0.5	17.3	0.4	20.4	Cyprus
0.2	22.9	0.4	1.8	Czech Republic
4.6	22.4	0.9	12.5	Denmark
0.9	18.1	0.4	3.9	England
	17.4	0.3	11.3	France
2.2	13.0	1.6	14.9	Germany
3.2	15.3	0.8	13.0	Greece
0.8	25.7		11.0	Hong Kong
6.1	10.9	1.5	17.3	Hungary
1.5	23.8	1.9	6.3	Iceland
0.2	22.2	0.4	8.0	Iran, Islamic Republic
1.0	27.8		2.0	Ireland
1.4	3.3	0.6	8.8	Israel
	20.6	0.7	4.5	Japan
2.9	11.1		2.4	Korea
3.1	14.8	1.7	22.9	Kuwait
1.0	15.8	2.0	13.1	Latvia (LSS)
0.6	17.7	0.3	17.4	Lithuania
	36.4	0.4	1.5	Netherlands
2.1	22.2	0.5	4.5	New Zealand
	22.1	1.8	4.1	Norway
16.6	30.4	1.8	12.5	Philippines
3.0	33.8	0.2	9.3	Portugal
1.1	15.2	0.6	23.2	Romania
1.3	27.4	1.8	8.7	Russian Federation
1.1	34.6		5.5	Scotland
2.5	8.8	0.4	2.1	Singapore
1.2	12.3	0.5	4.9	Slovak Republic
1.8	21.5	1.0	5.2	Slovenia
9.2	30.8	1.2	34.4	South Africa
2.8	17.4		7.3	Spain
1.7	18.6	1.0	3.5	Sweden
1.9	20.8	0.5	6.4	Switzerland
8.8	14.3	0.0	6.9	Thailand

- A test is not a test. Comparative performances on the TIMSS achievement tests in mathematics and the sciences by students from different countries varied with the content of the test. Despite highly aggregated scores for 'mathematics' or 'science', achievement was not monolithic and was not consistent among content areas. The content on which students were tested mattered to how a country fared in comparison to

Exhibit 5.6. Fourth Grade Results for an Extended Response Science Item About Snow on the Mountain Top.

	Correct			Incorrect	
	10 Mentions that it is colder on the mountain tops or warmer farther down	11 Mentions that more snow is falling high up in the mountains	19 Other correct	70 Mentions that there is sunshine lower down or there is more sunshine lower down	71 Refers to sun or heat melting the snow
United States	49.6	2.5	0.5	2.4	9.6
International	40.6	1.8	3.4	3.3	10.1
Australia	34.0	1.3	2.0	5.6	18.6
Austria	27.9	0.2	2.3	2.2	4.0
Canada	53.2	0.2	0.7	2.5	7.8
Cyprus	22.0	1.8	5.3	1.7	14.1
Czech Republic	54.0	2.4	3.7	5.2	5.6
England	50.5	1.9	1.4	4.4	12.8
Greece	22.3	1.9	3.6	1.5	22.8
Hong Kong	43.8	1.0	1.5	1.0	2.0
Hungary	63.6	1.0	2.5	5.7	1.3
Iceland	45.4	1.3	3.5	4.9	5.6
Iran, Islamic Republic	26.1	0.6	15.7	4.2	3.2
Ireland	38.3	1.7	0.7	6.0	13.7
Israel	40.4	1.4	3.8	1.0	10.7
Japan	63.8	4.4	4.8	4.1	7.2
Korea	62.8	1.2	6.2	6.5	0.3
Kuwait	8.3	1.5	12.0		
Latvia (Lss)	42.7	1.1	3.4	1.4	5.9
Netherlands	49.7	1.6	1.2	2.5	21.9
New Zealand	36.7	2.6	1.5	3.3	25.4
Norway	62.2	1.8	0.3	3.1	1.8
Portugal	24.1	4.1	5.3	1.5	8.4
Scotland	43.1	3.3	1.7	4.1	12.8
Singapore	31.4	0.8	0.5	2.8	16.5
Slovenia	43.3	1.2	2.6	3.7	9.0
Thailand	16.7	4.2	2.8	1.6	12.0

> 09 Sometimes mountains can still have snow on their tops when the snow on the lower parts of the mountains has melted. What makes this happen?

others. Countries had profiles of contents on which they did better and on which they did worse, comparatively.

- No country could do it all and do it all well. With rare exceptions, most TIMSS countries performed among the top in at least some topic areas and many of these same countries performed far worse in other areas.

- While it is tempting to blame these topic differences on students, this explanation of differences among topics is not plausible. The same children took the test in all areas. Even when every child in a country did not take all items, there was sufficient overlap to allow estimates that clearly revealed differences among national performances for

Exhibit 5.6. (Continued).

Incorrect				Nonresponce		
72 Refers to the mountain being very high	73 Refers to the wind blowing more on the mountain top	76 Merely repeats information in stem	79 Other incorrect	90 Crossed out/ erased, illegible, or impossible to interpret	99 Blank	
4.4	**0.8**	**0.9**	**19.0**	**2.1**	**8.1**	United States
3.4	1.0	1.5	19.4	1.9	15.1	International
3.0	2.3	0.5	22.0	0.8	9.9	Australia
3.3	0.3	0.1	45.0	1.2	13.4	Austria
3.1	0.3	0.8	16.0	5.0	10.3	Canada
3.0	1.7	0.7	24.2	1.8	23.7	Cyprus
5.5	0.5	2.6	10.7	1.9	7.9	Czech Republic
2.4	0.5	2.1	9.8	5.0	9.2	England
4.0	0.1	1.1	19.5	0.9	22.2	Greece
2.8	0.1	3.4	24.0	3.4	17.0	Hong Kong
2.9		2.4	8.5	1.3	11.0	Hungary
1.5		1.2	17.1	3.7	15.8	Iceland
3.8		0.4	19.5	1.9	24.5	Iran, Islamic Republic
7.1	0.5	1.8	18.9	2.2	9.3	Ireland
3.7	0.4	3.0	17.5	3.0	15.1	Israel
1.3	2.8	0.3	8.3	0.2	2.9	Japan
0.2	3.6	1.2	17.3	0.8		Korea
		0.7	5.1		72.3	Kuwait
2.3	2.3	0.2	24.1	0.3	16.3	Latvia (Lss)
1.0	0.7	1.7	15.5	0.4	3.8	Netherlands
2.4		1.1	12.7	1.1	13.2	New Zealand
7.7		0.7	11.0	1.5	9.8	Norway
6.5	0.7	1.2	24.4	2.7	21.2	Portugal
5.1	0.6	2.5	17.2	0.6	9.1	Scotland
1.0	0.5	2.3	37.4	1.8	5.0	Singapore
4.3		3.6	17.2	2.8	12.4	Slovenia
1.8	0.4	3.8	42.6	0.5	13.7	Thailand

topics that did not depend on the students. This seems clearly to suggest that achievement differences may relate to curriculum differences and, to each country's goals, plans, emphases, and classroom activities.

- Previous TIMSS data made it clear that there were major curricular differences in mathematics and the sciences among participating countries.[5]

- Analyses are needed that formally link topic differences in achievement with curricular differences. These analyses have been done and have been reported elsewhere.[6] Some conclusions were, however, noted here.

- The US student achievements reflect the general point that achievement was not monolithic. Our results showed that we had both relatively strong and relatively weak areas in comparison with other TIMSS countries, both for mathematics and science and for nine- and thirteen-year-olds. Some strengths and weaknesses were the same for both fourth and eighth graders. Others were not, at least in mathematics.

- Portraying comparative strengths and weaknesses at the sub-topic and at the item levels suggests that the picture of the US' TIMSS performance in the context of other countries was more varied and complex than was suggested by the highly aggregated, global scores. We did not do all things equally well or equally poorly. If this relates to our curricula, these differences are important to deliberate efforts at change, reform, and improvement since they help reveal what worked and what did not.

- Performance assessment results further highlight the points above because they suggest still another dimension on which country performance varied – US performance in particular. There was considerable variation between more traditional written items and active, performance tasks even in the same areas, especially in science at fourth grade. There was also variation among the results for the different performance tasks themselves.

- Part of the bottom line on US comparative achievement thus far is that it demonstrates the earlier point that 'you can't do it all.' US curricula in mathematics through eighth grade, and for many grades in science, seems to be built around the idea of trying to do something for all topics at almost every grade. Our results for nine- and thirteen-year-olds make it clear that this does not work. We did well only in a few areas and those were not consistently the same across grades.

ASSESSING THE ACHIEVEMENT GAINS OF US STUDENTS

One of the strengths of the TIMSS design is that, while a true longitudinal data collection for a school year was not possible given other goals of the study, the possibility of quasi-longitudinal analyses were built into the design.[7] Students were tested in the two adjacent grades that contained the most nine-year-olds and in the two that contained the most thirteen-year-olds (as well as students at the end of secondary school). The samples were larger in some countries for the upper of the two grades, but were sufficiently large at the lower to allow national achievement estimates at both.

This yields pairs of achievement estimates ('cross-sections') approximately one grade apart. These can be treated at least as 'quasi-longitudinal' as long as we realize that we are not tracking precisely the same students and if we make a few assumptions. First, we must assume that there were no major cohort differences between those in pairs of grades – for example, that US seventh graders were not in some significant way qualitatively different from US eighth graders other than having not yet experienced eighth grade. Sampling was designed to make cohorts at both grades in each pair nationally representative of students at that grade level. This makes it unlikely that significant cohort differences existed. Second, we must ignore retention differences among countries over the two grades since students who were not retained might have differed significantly from those who were in how they achieved. This is not a significant problem in the US or in most of the countries because both nine- and thirteen-year-olds were below the age at which mandatory schooling ended.

With these assumptions, we can treat the grade pairs as allowing a longitudinal-like analysis of achievement gains. In most of the sample, the adjacent grades were in the same school, which makes even more reasonable the assumption that their comparison represents a grade's gain. We focus on the upper of the two grades (in the US, fourth and eighth grades), one from each pair (in the US third and fourth grades and seventh and eighth grades). These students were tested near the end of the school year and had had most of their grade's experience and access to learning possibilities. The differences between estimated national achievements at the lower and the upper of the two grades could be assumed (given the other assumptions above) to be attribute to the effects of schooling at the upper grade. Measures of achievement gains at the national level could thus be linked to aggregate national curriculum and educational system characteristics. For a few items, data was gathered at all four grade levels which allowed further comparisons. Of course, such wider time differences between grade cohorts make more problematic assumptions of cohort similarity, not in the demographic characteristics of students but in whether the curriculum had been different for those in the upper of each grade pair than they were for those in the lower of each grade pair the time of testing. Such comparisons are used sparingly.

These quasi-longitudinal possibilities allowed us to focus on what US students learned at fourth and eighth grades. Overwhelmingly, the answer was that little new was mastered in the US at either of those two grades. In light of that, most of US students' content mastery appears to have been the cumulative effect of small gains over many grades.

Content Area Gains. Across countries at the topic level, there were only seven countries out of 41 at the seventh and eighth grade level that were not in the top ten gaining countries for at least one mathematics topic. The US was one of those seven. The largest gain for the US in rankings among countries was in the area of 'estimating quantity and size', a content area for which we still ranked seventeenth among the 41 countries in eighth grade. This topic was obviously more prevalent in US eighth grade instruction than in previous grades. This was among the areas in which we performed most poorly at fourth grade, but by the end of eighth grade we did not differ significantly from the international average.

Exhibit 5.7. *Percentage Points Gain in Specific Mathematics Content Areas for Fourth Grade Students (national mean percent gain).*

Meaning of Whole Numbers

Norway	18.7	(1.2)
Iceland	17.2	(1.5)
Latvia	16.5	(1.4)
Cyprus	16.5	(1.1)
Netherlands	16.4	(1.1)
Austria	15.2	(1.2)
Canada	14.7	(1.3)
New Zealand	14.5	(1.5)
Hungary	14.4	(1.3)
Greece	14.1	(1.5)
Czech Republic	14.1	(1.0)
International	*13.1*	
Australia	13.1	(1.3)
Scotland	13.0	(1.3)
Ireland	13.0	(1.2)
England	12.4	(1.1)
Iran	12.3	(1.5)
Slovenia	12.1	(1.1)
USA	**12.1**	**(1.0)**
Portugal	11.5	(1.4)
Thailand	9.6	(1.8)
Hong Kong	9.6	(1.2)
Japan	9.0	(0.7)
Singapore	7.5	(1.0)
Korea	6.6	(0.7)

Whole Number Operations

Norway	24.3	(1.4)
Ireland	20.4	(1.3)
Netherlands	20.4	(1.2)
Iceland	20.1	(1.5)
Latvia	18.8	(1.7)
Austria	18.5	(1.5)
Greece	18.2	(1.7)
Scotland	17.9	(1.4)
Czech Republic	17.8	(1.2)
Slovenia	17.4	(1.4)
Cyprus	17.0	(1.3)
USA	**16.5**	**(1.3)**
International	*15.9*	
Canada	15.5	(1.5)
New Zealand	15.4	(1.5)
Hungary	14.8	(1.3)
Australia	14.7	(1.3)
England	13.1	(1.3)
Iran	12.4	(1.8)
Japan	12.1	(0.8)
Hong Kong	12.1	(1.3)
Thailand	11.8	(2.0)
Portugal	11.0	(1.5)
Korea	10.8	(0.8)
Singapore	9.6	(1.2)

Common Fractions

Netherlands	23.3	(1.5)
Austria	17.8	(1.5)
Hungary	16.6	(1.7)
Czech Republic	16.3	(1.3)
Ireland	15.2	(1.6)
Hong Kong	15.1	(1.2)
Canada	15.0	(1.4)
Latvia	14.4	(1.5)
Slovenia	14.1	(1.2)
Scotland	13.2	(1.3)
Japan	13.0	(0.8)
International	*12.9*	
USA	**12.2**	**(1.4)**
Norway	12.1	(1.2)
Cyprus	12.0	(1.2)
England	11.7	(1.2)
Greece	11.2	(1.6)
Iran	11.2	(1.5)
Thailand	11.2	(1.7)
Australia	11.1	(1.5)
New Zealand	10.8	(1.6)
Singapore	10.4	(1.5)
Korea	9.0	(1.2)
Iceland	6.9	(1.3)
Portugal	5.8	(1.0)

Decimal Fractions

Singapore	36.8	(1.0)
Hong Kong	29.1	(1.6)
Cyprus	25.4	(1.4)
Iran	21.8	(1.8)
Ireland	21.8	(2.1)
Netherlands	19.8	(1.4)
USA	**18.6**	**(1.7)**
Australia	18.5	(1.5)
Scotland	16.9	(1.7)
International	*16.3*	
Austria	15.4	(2.0)
Greece	15.2	(2.1)
Canada	14.4	(1.4)
Latvia	14.4	(2.1)
Norway	14.2	(1.2)
Thailand	13.7	(2.0)
Japan	12.5	(1.2)
Portugal	11.8	(1.6)
Korea	11.6	(1.2)
Slovenia	11.3	(1.6)
Iceland	10.8	(1.5)
England	9.8	(1.4)
New Zealand	9.8	(1.6)
Hungary	9.7	(1.4)
Czech Republic	8.8	(1.4)

Perimeter, Area, & Volume

Norway	17.3	(1.6)
Slovenia	16.8	(1.4)
Singapore	16.2	(1.4)
Hungary	15.1	(1.5)
Netherlands	13.9	(1.5)
Greece	13.8	(1.7)
Cyprus	13.7	(1.6)
USA	**13.0**	**(1.2)**
Ireland	12.9	(1.4)
Austria	12.8	(1.9)
Czech Republic	12.1	(1.4)
International	*11.1*	
Portugal	10.6	(1.7)
England	10.2	(1.4)
Australia	10.2	(1.5)
Latvia	9.9	(1.9)
Iran	9.4	(1.6)
Scotland	9.4	(1.6)
Thailand	9.3	(1.7)
Canada	9.0	(1.7)
Hong Kong	8.5	(1.6)
Iceland	7.8	(1.8)
New Zealand	6.0	(1.6)
Japan	5.4	(0.9)
Korea	2.9	(1.2)

Geometry: Position & Shapes

Norway	11.8	(1.4)
Iceland	11.5	(1.7)
Ireland	11.0	(1.4)
Netherlands	10.6	(1.2)
Japan	10.4	(1.1)
Greece	10.0	(1.7)
Hungary	10.0	(1.5)
Canada	9.8	(1.4)
Portugal	9.7	(1.5)
England	9.5	(1.3)
Latvia	9.4	(1.5)
Austria	9.3	(1.6)
Australia	9.2	(1.4)
International	*9.0*	
Iran	8.9	(1.3)
Hong Kong	8.8	(1.2)
Singapore	8.6	(1.3)
USA	**8.3**	**(1.2)**
Slovenia	8.1	(1.2)
Czech Republic	8.0	(1.2)
Thailand	7.6	(1.9)
New Zealand	7.6	(1.6)
Cyprus	7.1	(1.3)
Scotland	6.0	(1.3)
Korea	3.9	(1.1)

Symmetry, Congruence, & Similarity

Cyprus	21.0	(2.2)
Hungary	19.7	(1.8)
Singapore	19.4	(1.2)
Norway	18.0	(2.1)
Iceland	16.2	(2.4)
Austria	15.4	(2.2)
Greece	15.3	(2.3)
Ireland	14.8	(1.7)
Thailand	12.7	(2.9)
USA	**12.6**	**(1.6)**
International	*12.6*	
Czech Republic	12.2	(1.6)
England	12.0	(1.6)
Netherlands	11.2	(1.8)
Latvia	10.8	(2.5)
Portugal	10.4	(2.1)
Japan	10.1	(1.3)
Canada	9.8	(2.2)
Australia	9.7	(2.2)
Iran	9.7	(2.1)
New Zealand	9.6	(2.7)
Scotland	9.5	(1.8)
Hong Kong	9.1	(1.5)
Korea	7.7	(1.3)
Slovenia	6.2	(1.8)

Proportionality

Netherlands	19.1	(1.4)
Hungary	17.7	(1.7)
Czech Republic	17.6	(1.5)
Norway	16.8	(1.3)
Singapore	15.2	(1.9)
Austria	15.2	(1.5)
Ireland	15.1	(1.6)
Slovenia	14.8	(1.5)
Korea	14.3	(1.1)
Japan	14.1	(0.9)
Cyprus	13.4	(1.1)
International	*13.1*	
Iceland	13.0	(1.6)
Scotland	12.7	(1.4)
USA	**12.5**	**(1.1)**
Canada	12.4	(1.5)
New Zealand	12.3	(1.6)
Australia	12.3	(1.4)
Hong Kong	11.9	(1.5)
England	11.1	(1.1)
Latvia	10.8	(1.8)
Greece	9.5	(1.5)
Portugal	8.8	(1.1)
Thailand	7.9	(1.7)
Iran	6.0	(1.0)

Exhibit 5.7. (Continued).

Estimating Quantity & Size		
Hungary	18.1	(1.8)
Hong Kong	15.3	(1.5)
Slovenia	15.1	(1.5)
Norway	14.6	(1.4)
Czech Republic	14.4	(1.2)
Austria	13.2	(1.6)
Ireland	13.1	(1.5)
Portugal	13.0	(1.6)
Netherlands	12.7	(1.3)
Latvia	11.5	(2.0)
International	*11.5*	
Cyprus	11.4	(1.2)
Greece	11.2	(1.7)
New Zealand	10.8	(1.6)
Iceland	10.7	(1.5)
Australia	10.3	(1.4)
Korea	10.0	(1.2)
Thailand	9.8	(1.8)
Singapore	9.5	(1.5)
Canada	9.4	(1.2)
Scotland	9.1	(1.5)
USA	**8.2**	**(1.3)**
Iran	8.1	(1.6)
Japan	7.9	(0.9)
England	7.5	(1.3)

Rounding & Estimating Computations		
Japan	35.6	(1.5)
Singapore	26.3	(1.6)
Hong Kong	24.3	(1.8)
Austria	23.8	(2.6)
Netherlands	22.6	(1.8)
Canada	22.3	(2.2)
Ireland	21.8	(1.7)
Czech Republic	21.4	(1.9)
Slovenia	20.0	(1.7)
Hungary	19.5	(1.9)
USA	**18.6**	**(1.7)**
International	*17.0*	
Korea	16.9	(1.5)
Norway	16.5	(1.8)
Australia	13.8	(1.7)
Scotland	13.5	(1.9)
New Zealand	13.3	(2.1)
Iceland	12.8	(2.1)
Greece	12.6	(2.2)
Cyprus	12.2	(1.8)
Latvia	12.1	(2.2)
England	10.6	(1.9)
Portugal	8.7	(2.1)
Thailand	4.9	(2.1)
Iran	4.7	(1.5)

Measurement Units		
Norway	19.7	(1.5)
Hungary	18.7	(1.5)
Slovenia	16.2	(1.5)
Netherlands	15.6	(1.2)
Portugal	15.4	(1.3)
Ireland	15.0	(1.3)
Cyprus	14.7	(1.2)
Austria	14.0	(1.4)
Iceland	13.9	(1.2)
Czech Republic	13.9	(1.2)
Latvia	13.6	(1.7)
Greece	13.3	(1.6)
International	*13.2*	
Hong Kong	12.7	(1.4)
Scotland	12.7	(1.5)
Australia	12.4	(1.2)
Korea	11.9	(1.0)
New Zealand	11.2	(1.5)
USA	**11.2**	**(1.1)**
Canada	11.0	(1.3)
Japan	10.4	(0.7)
England	10.1	(1.1)
Singapore	9.9	(1.6)
Thailand	9.6	(1.8)
Iran	8.8	(1.6)

Patterns, Relations & Functions		
Iceland	17.3	(2.0)
Norway	17.2	(1.7)
Slovenia	15.2	(1.6)
Cyprus	15.0	(1.6)
Austria	14.6	(1.8)
Canada	14.3	(1.7)
Czech Republic	13.9	(1.5)
Ireland	13.3	(1.6)
Hungary	13.2	(1.8)
Australia	13.1	(1.4)
Netherlands	12.8	(1.5)
International	*12.7*	
USA	**12.6**	**(1.4)**
Portugal	12.2	(1.8)
Scotland	11.9	(1.3)
Greece	11.6	(1.7)
Thailand	11.6	(1.9)
England	11.6	(1.3)
New Zealand	11.5	(1.7)
Hong Kong	11.4	(1.5)
Latvia	11.3	(2.0)
Japan	10.8	(0.9)
Korea	10.4	(1.0)
Singapore	8.6	(1.4)
Iran	8.2	(1.7)

Equations & Formulas		
Norway	21.4	(2.0)
Netherlands	20.9	(2.2)
Iceland	20.4	(2.5)
Austria	20.2	(2.6)
Ireland	18.7	(2.0)
Greece	17.5	(2.4)
Canada	17.1	(2.1)
Australia	16.9	(2.3)
USA	**16.6**	**(1.8)**
New Zealand	16.6	(2.2)
Cyprus	16.5	(1.8)
Latvia	16.1	(2.2)
Hong Kong	15.9	(1.7)
Hungary	15.7	(2.0)
Scotland	15.5	(2.0)
International	*15.5*	
England	15.5	(2.1)
Czech Republic	14.0	(1.8)
Iran	12.6	(2.6)
Japan	11.9	(1.1)
Singapore	11.8	(1.4)
Portugal	11.7	(2.4)
Slovenia	11.7	(2.0)
Thailand	10.4	(2.8)
Korea	6.1	(1.3)

Data Representation Probability & Statistics		
Norway	22.7	(1.4)
Iceland	20.1	(1.6)
Netherlands	19.4	(1.4)
Ireland	19.3	(1.5)
Cyprus	19.1	(1.2)
Austria	18.3	(1.9)
New Zealand	18.0	(1.8)
Scotland	17.5	(1.5)
USA	**17.2**	**(1.3)**
Czech Republic	16.1	(1.4)
Australia	16.1	(1.5)
Hungary	15.8	(1.5)
Greece	15.6	(1.6)
International	*15.4*	
Canada	15.3	(1.7)
Thailand	14.6	(2.3)
Latvia	14.0	(1.7)
England	13.1	(1.4)
Hong Kong	12.8	(1.4)
Singapore	12.7	(1.4)
Portugal	12.3	(1.5)
Slovenia	12.1	(1.4)
Korea	10.1	(1.0)
Japan	10.0	(0.7)
Iran	6.9	(1.2)

Exhibit 5.8. Percentage Points Gain in Specific Science Content Areas for Fourth Grade Students (national mean percent gain).

Earth Features		
Slovenia	14.3	(1.0)
Hungary	13.8	(1.3)
Norway	13.6	(1.1)
Netherlands	12.5	(1.0)
Ireland	11.8	(1.2)
Iceland	11.7	(1.3)
Canada	11.2	(1.1)
Portugal	10.6	(1.1)
New Zealand	10.5	(1.4)
Czech Republic	10.2	(1.1)
International	*10.2*	
Latvia	10.0	(1.7)
Hong Kong	9.9	(1.1)
England	9.9	(1.0)
Austria	9.9	(1.4)
USA	**9.5**	**(1.2)**
Greece	9.5	(1.5)
Scotland	9.2	(1.2)
Cyprus	9.1	(1.1)
Korea	8.4	(0.9)
Singapore	8.3	(1.2)
Iran	7.9	(1.1)
Japan	7.7	(0.7)
Australia	7.3	(1.2)
Thailand	7.0	(1.8)

Earth Processes		
Norway	14.1	(1.4)
Greece	13.4	(1.7)
Hungary	13.2	(1.5)
Slovenia	12.7	(1.6)
Netherlands	12.1	(1.4)
Latvia	11.5	(1.9)
Portugal	10.9	(1.6)
Czech Republic	10.7	(1.5)
Ireland	10.7	(1.4)
Japan	10.5	(1.4)
Cyprus	10.4	(1.2)
Canada	10.4	(1.7)
Singapore	10.0	(1.2)
International	*9.9*	
Iceland	9.7	(1.8)
USA	**9.4**	**(1.5)**
New Zealand	8.7	(1.8)
Thailand	8.4	(2.1)
Scotland	8.2	(1.4)
Iran	7.9	(1.4)
Hong Kong	7.9	(1.1)
England	7.6	(1.3)
Austria	6.6	(1.8)
Korea	6.5	(1.2)
Australia	6.2	(1.3)

Earth in the Universe		
Hungary	18.3	(1.5)
Czech Republic	13.6	(1.3)
Norway	13.5	(1.5)
Netherlands	11.3	(1.5)
Australia	9.3	(1.5)
Iceland	9.0	(1.6)
Canada	8.8	(1.5)
Austria	8.5	(1.6)
Slovenia	8.5	(1.4)
Hong Kong	8.5	(1.1)
New Zealand	8.2	(1.8)
Portugal	8.1	(1.7)
International	*8.1*	
Iran	7.8	(1.8)
Thailand	7.5	(2.0)
Japan	7.0	(1.1)
USA	**6.9**	**(1.2)**
Korea	6.5	(1.2)
Ireland	6.4	(1.4)
Scotland	6.1	(1.4)
England	4.8	(1.4)
Singapore	4.6	(1.4)
Latvia	4.4	(1.7)
Greece	4.3	(1.5)
Cyprus	2.5	(1.3)

Plants & Animals		
Cyprus	14.2	(1.2)
Iceland	14.1	(1.9)
Norway	13.8	(1.5)
Hungary	13.3	(1.4)
Czech Republic	11.2	(1.3)
Netherlands	11.0	(1.3)
Iran	10.9	(1.2)
Slovenia	10.7	(1.3)
Greece	10.1	(1.6)
Australia	9.4	(1.3)
International	*9.0*	
England	8.7	(1.2)
Canada	8.3	(1.3)
Ireland	8.2	(1.2)
New Zealand	7.8	(1.8)
USA	**7.5**	**(1.3)**
Hong Kong	6.9	(1.1)
Scotland	6.9	(1.3)
Austria	6.9	(1.6)
Latvia	6.5	(1.6)
Korea	6.5	(1.0)
Thailand	6.2	(1.4)
Portugal	5.6	(1.5)
Japan	5.4	(0.8)
Singapore	4.8	(1.2)

Human Biology & Health		
Norway	15.3	(1.2)
Singapore	14.8	(1.4)
Austria	13.4	(1.2)
Iran	13.3	(1.2)
Iceland	11.7	(1.2)
Hungary	11.5	(1.0)
Hong Kong	11.3	(1.1)
Cyprus	11.0	(0.9)
Canada	10.8	(0.9)
Japan	10.6	(0.6)
Ireland	10.6	(1.1)
International	*10.3*	
Australia	10.2	(1.0)
USA	**9.8**	**(0.9)**
Scotland	9.7	(1.2)
New Zealand	9.7	(1.5)
Netherlands	9.7	(0.9)
Portugal	9.3	(1.1)
Greece	9.3	(1.2)
Czech Republic	8.8	(0.8)
England	8.5	(0.9)
Thailand	7.9	(1.9)
Latvia	7.6	(1.3)
Slovenia	7.3	(1.1)
Korea	5.4	(0.7)

Matter		
Netherlands	14.1	(1.3)
Singapore	13.5	(1.4)
Austria	13.5	(1.7)
Norway	13.1	(1.4)
Iceland	12.6	(1.4)
Hungary	12.5	(1.4)
Cyprus	11.6	(1.1)
Czech Republic	11.4	(1.2)
Canada	10.8	(1.2)
Slovenia	10.6	(1.4)
New Zealand	10.5	(1.7)
Australia	10.4	(1.2)
International	*10.4*	
Iran	10.1	(1.4)
Scotland	10.1	(1.3)
USA	**9.9**	**(1.1)**
Ireland	9.6	(1.2)
Latvia	9.2	(1.7)
Korea	9.1	(0.9)
Japan	8.7	(0.8)
Hong Kong	8.5	(1.3)
Thailand	8.3	(1.8)
Portugal	7.5	(1.4)
Greece	7.1	(1.5)
England	7.0	(1.2)

Energy and Physical Processes		
Norway	13.9	(1.4)
Czech Republic	12.5	(1.1)
Netherlands	11.8	(1.3)
Iceland	11.4	(1.5)
Canada	10.9	(1.2)
Austria	10.8	(1.6)
Hungary	10.7	(1.3)
Australia	10.3	(1.4)
International	*9.6*	
Portugal	9.6	(1.5)
Iran	9.6	(1.3)
Slovenia	9.4	(1.1)
Cyprus	9.4	(1.2)
USA	**9.2**	**(1.1)**
Ireland	9.2	(1.2)
England	9.2	(1.2)
Hong Kong	9.2	(1.3)
Greece	9.1	(1.6)
Scotland	8.9	(1.3)
Singapore	8.8	(1.2)
New Zealand	8.6	(1.6)
Japan	8.6	(0.8)
Korea	8.0	(0.9)
Latvia	7.0	(1.4)
Thailand	4.6	(1.7)

Physical & Chemical Changes		
Singapore	18.6	(1.4)
Norway	16.0	(1.8)
Cyprus	15.4	(1.5)
Netherlands	15.3	(1.7)
Austria	14.8	(1.8)
Iceland	13.7	(1.8)
Portugal	13.5	(1.7)
Ireland	13.3	(1.6)
Canada	13.2	(1.4)
Hong Kong	13.0	(1.6)
Greece	12.7	(2.0)
Iran	12.3	(2.0)
Japan	12.3	(1.2)
International	*12.2*	
Slovenia	12.1	(1.7)
New Zealand	12.0	(2.2)
Australia	11.7	(1.6)
Scotland	11.5	(1.6)
Hungary	11.1	(1.7)
England	11.0	(1.6)
USA	**9.9**	**(1.5)**
Czech Republic	8.9	(1.5)
Thailand	8.5	(2.6)
Latvia	7.7	(2.1)
Korea	3.6	(1.3)

Exhibit 5.8. *(Continued).*

Organs & Tissues		
Singapore	20.3	(1.4)
Iran	14.7	(1.3)
Austria	12.4	(1.4)
Norway	12.0	(1.5)
Hong Kong	11.8	(1.2)
Cyprus	11.5	(1.2)
Hungary	10.8	(1.3)
Japan	10.6	(0.9)
Canada	10.2	(1.0)
International	*9.7*	
Greece	9.4	(1.6)
USA	**9.3**	**(1.2)**
Ireland	9.3	(1.1)
Scotland	9.2	(1.5)
New Zealand	9.1	(1.5)
Netherlands	9.0	(1.1)
England	8.3	(1.3)
Czech Republic	8.3	(1.1)
Iceland	8.1	(1.8)
Australia	8.0	(1.4)
Thailand	7.8	(1.9)
Portugal	6.8	(1.5)
Latvia	5.7	(1.4)
Slovenia	5.0	(1.3)
Korea	4.8	(1.2)

Life Processes & Functions		
Norway	17.5	(1.6)
Iran	16.3	(1.7)
Iceland	15.4	(1.5)
Japan	15.3	(0.9)
Singapore	14.4	(1.5)
Cyprus	13.7	(1.0)
Greece	13.6	(1.4)
Hong Kong	13.4	(1.3)
Austria	12.9	(1.4)
Hungary	12.0	(1.3)
International	*11.8*	
Latvia	11.7	(1.8)
Scotland	11.7	(1.5)
Czech Republic	11.3	(1.2)
Portugal	11.2	(1.6)
Netherlands	10.3	(1.1)
Ireland	10.3	(1.3)
Canada	10.1	(1.6)
New Zealand	10.1	(1.8)
England	9.6	(1.3)
Slovenia	9.6	(1.4)
Australia	9.5	(1.3)
USA	**9.2**	**(1.2)**
Thailand	7.8	(2.1)
Korea	6.0	(1.1)

Life Cycles & Genetics		
Norway	13.2	(1.3)
Hungary	13.1	(1.3)
Iceland	12.3	(1.8)
Cyprus	11.3	(1.1)
Slovenia	9.8	(1.3)
Czech Republic	9.7	(1.0)
Australia	9.5	(1.1)
Greece	9.3	(1.5)
Hong Kong	8.9	(1.1)
Austria	8.6	(1.5)
International	*8.5*	
Ireland	8.4	(1.1)
USA	**8.3**	**(1.1)**
Netherlands	7.9	(1.0)
Singapore	7.7	(1.0)
England	7.7	(1.0)
Iran	7.5	(1.3)
Canada	7.3	(0.9)
Portugal	7.2	(1.4)
New Zealand	7.1	(1.5)
Scotland	6.5	(1.3)
Thailand	6.4	(1.9)
Korea	5.3	(0.8)
Latvia	5.3	(1.5)
Japan	5.0	(0.6)

Interactions of Living Things		
Norway	19.0	(1.6)
Iceland	16.8	(2.0)
Cyprus	14.3	(2.0)
Greece	13.3	(1.9)
Hong Kong	12.9	(1.6)
Iran	12.6	(2.0)
Slovenia	12.4	(1.6)
Latvia	12.2	(1.8)
Singapore	12.2	(1.6)
Czech Republic	12.1	(1.4)
Thailand	11.9	(2.0)
Portugal	11.5	(1.7)
International	*11.0*	
Ireland	10.5	(1.5)
New Zealand	10.3	(2.0)
Austria	9.7	(1.6)
Canada	9.5	(1.3)
USA	**9.5**	**(1.2)**
Australia	9.1	(1.4)
Netherlands	8.9	(1.4)
Scotland	8.7	(1.6)
England	8.3	(1.4)
Hungary	7.3	(1.8)
Japan	6.0	(0.8)
Korea	5.7	(1.2)

Forces and Motion		
Norway	16.2	(2.3)
Austria	15.0	(2.5)
Hungary	13.9	(1.9)
Singapore	13.3	(1.5)
Thailand	12.3	(2.2)
Netherlands	12.2	(1.7)
Canada	11.9	(1.7)
Ireland	11.7	(1.9)
Czech Republic	11.5	(1.7)
Slovenia	11.4	(1.9)
Hong Kong	10.9	(1.7)
New Zealand	10.2	(2.1)
International	*10.1*	
Greece	9.9	(2.4)
Iceland	9.4	(2.3)
Japan	8.9	(1.3)
Latvia	8.8	(2.6)
Iran	8.7	(1.9)
USA	**8.6**	**(1.5)**
Australia	7.9	(1.7)
Cyprus	7.9	(1.9)
Portugal	7.7	(1.9)
Scotland	5.3	(1.9)
Korea	4.6	(1.6)
England	4.2	(2.0)

Environmental & Resource Issues		
Norway	17.9	(1.3)
Iceland	15.8	(1.8)
Austria	15.1	(1.7)
Hungary	14.4	(1.3)
Czech Republic	13.5	(1.3)
Slovenia	13.5	(1.3)
Netherlands	13.2	(1.3)
Ireland	13.0	(1.2)
Australia	12.2	(1.4)
Portugal	12.1	(1.5)
Singapore	12.0	(1.6)
International	*11.9*	
Scotland	11.8	(1.5)
Canada	11.8	(1.0)
Hong Kong	11.1	(1.3)
Cyprus	10.9	(1.2)
New Zealand	10.8	(1.8)
USA	**10.7**	**(1.0)**
Japan	10.2	(0.8)
England	10.0	(1.4)
Greece	10.0	(1.5)
Iran	9.8	(1.1)
Thailand	9.3	(2.1)
Latvia	8.9	(1.7)
Korea	7.3	(1.1)

Scientific Processes		
Norway	16.0	(2.3)
USA	**15.7**	**(2.1)**
Iceland	15.5	(2.6)
Australia	15.4	(2.1)
Japan	13.1	(2.0)
Hong Kong	12.7	(2.0)
Netherlands	12.6	(2.2)
New Zealand	12.6	(2.8)
Singapore	12.5	(1.8)
Czech Republic	12.2	(2.3)
Austria	12.0	(2.1)
Canada	10.7	(2.4)
International	*10.7*	
Greece	10.4	(2.6)
England	10.3	(2.2)
Korea	9.9	(2.0)
Ireland	9.3	(1.9)
Slovenia	9.3	(2.0)
Cyprus	9.2	(2.0)
Scotland	9.1	(2.2)
Portugal	8.6	(2.0)
Hungary	7.9	(2.1)
Thailand	4.5	(3.1)
Iran	4.4	(2.2)
Latvia	4.0	(2.5)

Exhibit 5.9. *Percentage Points Gain in Specific Mathematics Content Areas for Eighth Grade Students (national mean percent gain).*

Whole Numbers & Common Fractions

Country	Gain	(SE)
France	10.5	(1.8)
Portugal	9.9	(1.5)
Lithuania	9.8	(2.2)
Czech Republic	9.6	(2.0)
Hungary	8.7	(1.7)
Spain	8.7	(1.6)
Slovak Republic	8.4	(1.7)
Sweden	8.4	(1.6)
Japan	7.3	(1.3)
Denmark	7.2	(2.0)
Latvia	7.2	(2.0)
Norway	6.8	(2.0)
Slovenia	6.7	(1.7)
England	6.6	(2.3)
Cyprus	6.5	(1.6)
Austria	6.1	(2.0)
Scotland	5.9	(2.2)
Greece	5.6	(1.6)
Singapore	5.5	(2.0)
New Zealand	5.4	(1.9)
International	*5.3*	
Hong Kong	5.2	(2.3)
Australia	5.0	(1.8)
Canada	4.8	(1.6)
Thailand	4.6	(2.1)
Ireland	4.5	(2.2)
Switzerland	4.4	(1.3)
Belgium (Fr)	4.2	(2.0)
Iceland	3.8	(2.7)
Russian Federation	3.7	(2.2)
Colombia	3.6	(2.5)
USA	3.4	(2.0)
Romania	3.3	(2.1)
Iran	3.1	(1.7)
Germany	2.9	(2.2)
Philippines	1.8	(1.8)
Bulgaria	1.8	(3.5)
South Africa	1.5	(1.9)
Netherlands	1.0	(3.3)
Korea	0.4	(1.7)
Belgium (Fl)	-0.3	(2.4)

Decimal Fractions & Percents

Country	Gain	(SE)
Spain	12.7	(1.0)
France	12.1	(1.3)
Sweden	10.0	(1.2)
Norway	9.2	(1.1)
Slovak Republic	8.6	(1.3)
Switzerland	8.6	(1.1)
Denmark	8.2	(1.2)
Slovenia	8.2	(1.3)
Lithuania	8.2	(1.5)
Czech Republic	8.1	(1.9)
New Zealand	7.1	(1.5)
Scotland	6.9	(1.8)
Iran	6.5	(1.1)
Latvia	6.5	(1.4)
Austria	6.5	(1.4)
Singapore	6.3	(1.7)
Russian Federation	6.2	(1.8)
Greece	6.1	(1.2)
Romania	6.0	(1.5)
International	*5.7*	
Iceland	5.6	(1.6)
Canada	5.6	(1.0)
Australia	5.1	(1.5)
Thailand	5.0	(2.2)
Hungary	5.0	(1.4)
Korea	4.9	(0.9)
USA	4.8	(1.9)
England	4.7	(1.3)
Japan	4.7	(0.7)
Belgium (Fr)	4.6	(1.5)
Bulgaria	3.9	(2.3)
Portugal	3.8	(1.0)
Hong Kong	3.5	(2.5)
Germany	3.4	(1.9)
Ireland	3.1	(1.7)
Colombia	2.5	(1.3)
Cyprus	2.1	(0.9)
Philippines	1.7	(1.3)
South Africa	0.4	(1.7)
Netherlands	0.2	(2.2)
Belgium (Fl)	-0.1	(1.6)

Relations of Fractions

Country	Gain	(SE)
Lithuania	11.4	(1.5)
Sweden	11.2	(1.1)
France	10.7	(1.4)
Hong Kong	9.3	(2.2)
Latvia	8.8	(1.3)
Czech Republic	8.7	(1.7)
Slovak Republic	8.5	(1.3)
Norway	7.9	(1.2)
Iceland	7.7	(1.7)
Hungary	7.7	(1.2)
Cyprus	7.0	(1.1)
Russian Federation	7.0	(1.4)
Slovenia	6.9	(1.2)
Portugal	6.7	(1.1)
Denmark	6.4	(1.3)
Switzerland	6.3	(1.3)
Canada	6.0	(1.0)
Romania	5.7	(1.3)
New Zealand	5.7	(1.4)
Scotland	5.4	(1.8)
International	*5.3*	
Greece	5.1	(1.1)
Netherlands	4.4	(2.1)
England	4.2	(1.5)
Australia	4.0	(1.3)
USA	3.9	(1.7)
Bulgaria	3.8	(2.3)
Korea	3.4	(1.0)
Austria	3.4	(1.3)
Japan	3.3	(0.7)
Germany	3.1	(1.3)
Thailand	3.0	(1.9)
Singapore	2.8	(1.5)
Iran	2.8	(1.0)
Spain	2.5	(0.9)
Belgium (Fr)	2.1	(1.5)
Ireland	2.1	(1.6)
Colombia	1.9	(1.4)
Philippines	1.7	(1.2)
Belgium (Fl)	-1.0	(1.7)
South Africa	-1.0	(1.8)

Estimating Quantity & Size

Country	Gain	(SE)
Sweden	13.5	(1.4)
Lithuania	11.0	(1.8)
Norway	10.5	(1.4)
Slovenia	10.1	(1.6)
France	9.7	(1.5)
Spain	9.5	(1.1)
Denmark	8.8	(1.5)
Czech Republic	8.3	(2.3)
Russian Federation	8.2	(2.0)
Hungary	8.1	(1.6)
Slovak Republic	7.9	(1.6)
New Zealand	7.4	(1.8)
Switzerland	7.1	(1.3)
Canada	7.1	(1.0)
Romania	6.8	(1.6)
Scotland	6.6	(2.0)
Austria	6.3	(1.3)
Greece	6.0	(1.4)
International	*5.7*	
England	5.3	(1.5)
Iceland	5.3	(1.9)
Singapore	5.3	(1.8)
Germany	5.2	(1.8)
Korea	5.2	(1.2)
Thailand	5.2	(2.1)
Australia	4.9	(1.5)
Latvia	4.7	(1.6)
USA	4.5	(1.8)
Belgium (Fr)	4.2	(1.8)
Japan	4.1	(0.9)
Hong Kong	4.1	(2.3)
Iran	3.7	(1.1)
Portugal	3.4	(1.4)
Bulgaria	3.0	(2.8)
Ireland	2.7	(1.8)
Cyprus	2.6	(1.2)
Philippines	2.6	(1.5)
Netherlands	1.4	(2.3)
Colombia	1.2	(1.3)
South Africa	0.4	(1.8)
Belgium (Fl)	-3.3	(2.3)

2-D Geometry

Country	Gain	(SE)
Switzerland	17.4	(1.6)
New Zealand	11.5	(1.8)
Greece	11.2	(1.4)
Lithuania	10.3	(2.0)
Norway	9.6	(1.7)
Ireland	9.6	(1.9)
Japan	8.4	(0.9)
Canada	8.2	(1.3)
Spain	8.1	(1.4)
Hungary	7.5	(1.8)
Czech Republic	7.3	(2.0)
France	7.0	(1.7)
Singapore	6.5	(1.6)
Denmark	6.5	(1.8)
Slovenia	6.1	(1.8)
Scotland	6.1	(2.0)
Russian Federation	5.9	(2.0)
International	*5.7*	
Portugal	5.7	(1.5)
Hong Kong	5.7	(2.6)
Slovak Republic	5.5	(1.5)
Australia	5.2	(1.6)
Latvia	5.2	(1.9)
Germany	5.0	(2.1)
Romania	4.8	(1.7)
USA	4.7	(1.9)
Sweden	4.7	(1.2)
Netherlands	3.9	(2.7)
Bulgaria	3.7	(3.0)
South Africa	3.5	(1.8)
Austria	3.5	(2.0)
Iceland	3.3	(3.0)
England	3.1	(1.9)
Iran	3.0	(1.8)
Belgium (Fl)	2.8	(2.2)
Colombia	2.0	(2.1)
Philippines	1.9	(1.3)
Cyprus	1.9	(1.4)
Korea	1.3	(1.4)
Belgium (Fr)	0.8	(2.0)
Thailand	0.4	(2.0)

Polygons & Circles

Country	Gain	(SE)
Lithuania	22.2	(2.0)
Latvia	17.2	(2.1)
Switzerland	14.2	(1.6)
Russian Federation	14.1	(2.8)
France	11.8	(2.0)
Japan	11.8	(1.3)
Hungary	11.0	(1.7)
Greece	10.4	(1.6)
Iceland	9.4	(2.0)
Slovenia	9.1	(1.8)
Denmark	9.0	(1.9)
Norway	8.7	(1.8)
England	8.4	(2.2)
Sweden	8.3	(1.6)
Canada	7.9	(1.5)
Netherlands	7.8	(2.8)
New Zealand	7.6	(1.9)
International	*7.6*	
Romania	7.5	(2.0)
Cyprus	7.4	(1.5)
Korea	7.3	(1.6)
Ireland	7.2	(2.2)
Czech Republic	7.1	(2.3)
Australia	6.4	(2.0)
Slovak Republic	6.2	(1.7)
Hong Kong	6.1	(3.1)
Germany	5.9	(2.6)
Singapore	5.7	(2.2)
Belgium (Fl)	5.2	(2.4)
Thailand	5.0	(2.0)
Bulgaria	5.0	(3.0)
Colombia	4.9	(1.9)
Belgium (Fr)	4.8	(2.0)
Scotland	4.7	(2.2)
USA	4.4	(1.7)
Austria	4.0	(2.3)
Portugal	3.4	(2.0)
Philippines	2.8	(1.3)
Iran	2.7	(2.0)
Spain	1.8	(1.6)
South Africa	0.1	(1.7)

3-D Geometry & Transformations

Country	Gain	(SE)
Switzerland	14.2	(1.5)
Greece	13.3	(1.4)
Lithuania	12.0	(2.0)
Ireland	9.8	(2.0)
New Zealand	9.5	(2.2)
Bulgaria	9.4	(2.3)
Belgium (Fr)	8.8	(1.8)
Norway	8.6	(1.7)
Singapore	8.4	(2.1)
Slovenia	8.0	(1.6)
Spain	7.6	(1.2)
Austria	7.5	(1.8)
Russian Federation	7.5	(1.7)
Portugal	7.2	(1.6)
Denmark	6.9	(1.7)
Hungary	6.8	(1.7)
Korea	6.8	(1.5)
England	6.6	(1.9)
Latvia	6.6	(1.6)
International	*6.6*	
USA	6.5	(1.9)
Slovak Republic	6.0	(1.6)
Netherlands	6.0	(2.4)
Canada	5.9	(1.3)
Germany	5.8	(2.1)
Sweden	5.7	(1.6)
Scotland	5.6	(2.2)
France	5.4	(1.5)
Czech Republic	5.3	(1.7)
Belgium (Fl)	5.3	(2.1)
Thailand	5.2	(1.9)
Australia	5.1	(1.7)
Cyprus	4.9	(1.6)
Philippines	4.2	(1.6)
Japan	4.2	(0.8)
South Africa	4.2	(1.9)
Iceland	2.8	(1.8)
Colombia	2.7	(1.9)
Iran	2.7	(2.3)
Hong Kong	2.2	(2.5)
Romania	1.1	(1.7)

Congruence & Similarity

Country	Gain	(SE)
Lithuania	20.8	(1.7)
Japan	16.1	(1.0)
Latvia	12.2	(1.6)
Russian Federation	11.5	(2.1)
Greece	11.5	(1.2)
Singapore	11.4	(2.1)
Slovak Republic	10.7	(1.5)
Czech Republic	10.1	(2.0)
Switzerland	10.0	(1.2)
Denmark	9.9	(1.6)
Belgium (Fl)	9.5	(1.9)
Norway	9.4	(1.4)
Hungary	9.3	(1.6)
Iran	8.8	(2.0)
Slovenia	8.6	(1.6)
Iceland	8.3	(2.0)
France	7.9	(1.6)
International	*7.5*	
Scotland	7.5	(2.0)
Thailand	7.4	(1.6)
Netherlands	7.2	(2.5)
Ireland	7.2	(1.8)
Portugal	7.0	(1.7)
Hong Kong	7.0	(2.7)
Canada	6.9	(1.1)
Australia	6.8	(1.3)
England	6.4	(1.8)
New Zealand	6.2	(1.6)
Korea	6.0	(1.2)
Sweden	5.7	(1.4)
Cyprus	5.5	(1.3)
Colombia	4.6	(1.5)
Romania	4.5	(1.9)
Germany	3.9	(2.1)
USA	3.5	(1.7)
Belgium (Fr)	3.5	(1.8)
Philippines	3.1	(1.3)
Spain	2.7	(1.4)
Austria	2.1	(1.7)
South Africa	2.1	(1.3)
Bulgaria	-1.3	(2.6)

Exhibit 5.9. *(Continued)*

Rounding		
Norway	9.4	(1.3)
Spain	9.4	(1.3)
Lithuania	8.8	(1.5)
Slovak Republic	8.5	(1.3)
England	8.5	(1.4)
Greece	8.3	(1.3)
Switzerland	8.2	(1.3)
Scotland	8.1	(2.0)
New Zealand	7.9	(1.5)
Denmark	7.8	(1.7)
Ireland	7.6	(1.7)
Sweden	7.0	(1.2)
Portugal	6.9	(1.4)
Canada	6.8	(1.1)
Slovenia	6.6	(1.2)
Cyprus	6.5	(1.2)
France	6.3	(1.4)
Latvia	6.3	(1.7)
Russian Federation	6.3	(1.9)
Czech Republic	6.2	(1.7)
USA	5.8	(1.9)
International	5.4	
Australia	5.4	(1.4)
Singapore	5.4	(1.4)
Romania	5.3	(1.4)
Bulgaria	5.2	(2.4)
Japan	4.5	(0.8)
Korea	4.5	(1.2)
Hungary	4.4	(1.5)
Austria	4.0	(1.7)
Iran	3.5	(2.1)
Hong Kong	3.3	(2.0)
Belgium (Fr)	3.3	(1.6)
Germany	3.1	(1.6)
Iceland	3.1	(1.7)
Netherlands	1.7	(1.7)
Thailand	1.5	(1.7)
Colombia	0.7	(1.4)
Philippines	0.6	(1.1)
Belgium (Fl)	-0.2	(1.2)
South Africa	-0.2	(1.7)

Estimating Computations		
Sweden	10.8	(1.5)
Lithuania	9.9	(2.1)
France	8.2	(1.4)
New Zealand	8.2	(1.6)
Latvia	8.1	(1.8)
Norway	6.4	(1.9)
Hungary	6.2	(1.6)
Scotland	6.1	(1.7)
Slovak Republic	5.9	(1.4)
Iceland	5.9	(1.9)
Greece	5.3	(1.5)
Hong Kong	5.1	(2.2)
Czech Republic	5.0	(1.4)
England	5.0	(1.9)
Cyprus	4.9	(1.8)
Portugal	4.8	(1.5)
Spain	4.7	(1.4)
Iran	4.4	(2.1)
Colombia	4.4	(1.9)
Russian Federation	4.2	(2.0)
Romania	4.0	(1.7)
International	4.0	
Denmark	3.4	(1.9)
Korea	2.9	(1.3)
Canada	2.9	(1.0)
Ireland	2.6	(1.6)
USA	2.5	(1.7)
Slovenia	2.3	(1.5)
Bulgaria	2.2	(3.1)
Philippines	1.7	(1.7)
Austria	1.7	(1.5)
Switzerland	1.7	(1.6)
Germany	1.6	(1.9)
Australia	1.5	(1.3)
Netherlands	1.5	(2.3)
Japan	1.3	(1.0)
Thailand	1.1	(1.9)
South Africa	0.6	(2.4)
Belgium (Fl)	0.4	(1.5)
Belgium (Fr)	-0.1	(1.9)
Singapore	-1.0	(1.1)

Measurement Units		
Lithuania	10.6	(1.9)
Sweden	10.2	(1.6)
Czech Republic	10.1	(2.1)
Iceland	9.9	(2.6)
Norway	9.4	(1.7)
Bulgaria	9.2	(3.1)
France	8.4	(1.8)
New Zealand	7.5	(2.0)
Hungary	7.4	(1.7)
Spain	7.4	(1.3)
Denmark	7.2	(1.9)
Russian Federation	7.2	(1.8)
Canada	6.9	(1.2)
Greece	6.8	(1.5)
Slovak Republic	6.5	(1.6)
Latvia	6.2	(1.7)
Hong Kong	6.1	(2.5)
England	5.8	(2.2)
International	5.7	
Slovenia	5.6	(1.6)
Switzerland	5.4	(1.5)
Cyprus	5.2	(1.5)
Korea	5.2	(1.6)
Ireland	5.1	(2.2)
Scotland	5.0	(2.1)
Japan	5.0	(1.1)
Colombia	5.0	(2.0)
Portugal	4.9	(1.5)
Singapore	4.9	(2.0)
Romania	4.7	(1.7)
Iran	4.6	(1.7)
USA	4.5	(2.2)
Thailand	4.5	(2.0)
Australia	4.2	(1.6)
Austria	3.7	(2.1)
Germany	3.6	(2.2)
Netherlands	2.0	(2.5)
Philippines	1.1	(1.9)
South Africa	0.4	(1.8)
Belgium (Fl)	0.4	(1.8)
Belgium (Fr)	-0.1	(2.1)

Proportionality Concepts		
Slovenia	10.7	(1.5)
Czech Republic	9.6	(2.0)
Hungary	9.2	(1.5)
Lithuania	8.7	(1.2)
Sweden	8.5	(1.4)
Russian Federation	8.4	(1.9)
Canada	8.2	(1.6)
Switzerland	7.8	(1.2)
Slovak Republic	7.7	(1.6)
France	7.6	(1.5)
Denmark	7.2	(1.4)
Latvia	7.2	(1.7)
Cyprus	7.0	(1.3)
Austria	6.9	(1.9)
Greece	6.8	(1.3)
Korea	6.8	(1.5)
Romania	6.4	(1.7)
Thailand	6.3	(2.1)
Australia	6.2	(1.5)
Japan	6.1	(1.0)
Iceland	6.1	(2.2)
Hong Kong	6.0	(2.7)
Singapore	6.0	(1.9)
New Zealand	5.7	(1.7)
International	5.7	
Norway	5.7	(1.3)
Scotland	5.5	(2.1)
Ireland	5.5	(1.9)
Spain	5.5	(1.2)
Portugal	4.4	(1.0)
England	4.2	(1.7)
USA	4.0	(1.9)
Germany	3.7	(1.7)
Belgium (Fr)	3.6	(1.7)
Netherlands	3.5	(2.5)
Colombia	2.2	(2.0)
Iran	1.7	(1.2)
Bulgaria	0.8	(3.1)
Belgium (Fl)	0.7	(2.0)
Philippines	0.6	(1.3)
South Africa	-0.6	(1.3)

Proportionality Problems		
Lithuania	10.9	(1.5)
Russian Federation	9.5	(1.8)
Czech Republic	9.0	(2.0)
Slovenia	8.5	(1.2)
Slovak Republic	8.2	(1.3)
France	7.8	(1.4)
Hungary	7.7	(1.3)
Hong Kong	7.3	(2.2)
Latvia	7.0	(1.3)
Korea	6.7	(1.1)
Australia	6.7	(1.3)
Japan	6.4	(0.8)
Canada	6.4	(1.0)
Portugal	6.4	(1.1)
Scotland	6.3	(1.8)
Switzerland	6.2	(1.0)
Romania	6.1	(1.5)
Denmark	5.9	(1.2)
Singapore	5.8	(1.9)
Cyprus	5.6	(0.9)
Sweden	5.6	(1.0)
Norway	5.5	(1.1)
International	5.3	
Thailand	5.0	(2.0)
Spain	4.8	(0.9)
Germany	4.5	(1.4)
Austria	4.3	(1.3)
New Zealand	4.3	(1.5)
Greece	4.2	(1.3)
Iran	3.9	(1.0)
USA	3.8	(1.7)
England	3.7	(1.6)
Belgium (Fr)	3.6	(1.5)
Iceland	3.0	(1.3)
Netherlands	2.9	(2.0)
Ireland	2.2	(1.6)
Philippines	2.0	(1.3)
Colombia	1.8	(1.3)
Bulgaria	1.8	(2.4)
South Africa	0.6	(1.3)
Belgium (Fl)	-1.5	(1.9)

Patterns, Relations & Functions		
Lithuania	9.5	(1.6)
Czech Republic	9.1	(1.7)
Portugal	8.7	(1.3)
Greece	8.6	(1.2)
Bulgaria	8.2	(3.3)
Switzerland	8.1	(1.2)
Slovenia	7.7	(1.3)
Iran	7.4	(1.4)
Hungary	7.3	(1.4)
Scotland	7.3	(1.7)
England	7.1	(1.7)
Russian Federation	7.1	(1.7)
Slovak Republic	6.9	(1.7)
New Zealand	6.8	(1.7)
Singapore	6.8	(2.0)
Spain	6.6	(1.3)
Norway	6.5	(1.5)
Cyprus	6.4	(1.3)
Austria	6.2	(1.4)
France	6.0	(1.3)
Canada	5.9	(1.1)
International	5.9	
Latvia	5.8	(1.7)
Iceland	5.7	(2.2)
Romania	5.6	(1.6)
Colombia	5.2	(2.5)
Thailand	5.2	(1.8)
Japan	5.1	(1.0)
Denmark	5.1	(1.6)
Sweden	4.8	(1.3)
Belgium (Fr)	4.7	(1.6)
Australia	4.7	(1.4)
Ireland	4.6	(1.7)
Korea	4.3	(1.4)
Netherlands	4.1	(2.1)
USA	3.9	(1.7)
Philippines	3.3	(1.4)
Hong Kong	3.1	(2.2)
Germany	2.4	(1.9)
South Africa	1.7	(1.8)
Belgium (Fl)	0.7	(1.9)

Exhibit 5.9. *(Continued)*

Perimeter, Area, & Volume

Country	Mean	(SE)
Iran	9.0	(1.7)
Lithuania	8.5	(1.7)
France	8.2	(1.6)
Sweden	7.2	(1.3)
Greece	6.9	(1.3)
Denmark	6.9	(1.6)
Iceland	6.7	(2.1)
Thailand	6.7	(2.0)
New Zealand	6.6	(1.7)
Russian Federation	6.3	(1.6)
Ireland	6.2	(1.7)
Czech Republic	6.1	(1.6)
Scotland	5.9	(1.9)
Switzerland	5.9	(1.2)
Canada	5.8	(1.2)
Spain	5.8	(1.1)
Slovenia	5.7	(1.2)
Japan	5.5	(1.0)
Australia	5.3	(1.3)
Romania	5.3	(1.6)
Slovak Republic	5.1	(1.3)
Latvia	5.1	(1.6)
International	*4.8*	
Norway	4.7	(1.4)
Portugal	4.6	(1.2)
Colombia	4.4	(1.9)
Hungary	4.3	(1.5)
Cyprus	4.2	(1.3)
England	4.2	(1.6)
Singapore	4.1	(1.9)
USA	**4.0**	**(2.0)**
Bulgaria	2.9	(2.6)
Hong Kong	2.9	(2.6)
Korea	2.5	(1.4)
Germany	2.3	(1.7)
Netherlands	2.1	(2.1)
Philippines	1.3	(1.3)
South Africa	1.2	(1.9)
Austria	0.9	(1.5)
Belgium (Fr)	0.3	(1.7)
Belgium (Fl)	-0.8	(1.7)

Measurement

Country	Mean	(SE)
Lithuania	12.6	(1.6)
Slovenia	12.3	(1.6)
Cyprus	11.7	(1.5)
Slovak Republic	11.6	(1.7)
Sweden	10.6	(1.5)
Denmark	10.6	(1.6)
France	10.5	(1.8)
Hungary	10.5	(1.6)
Czech Republic	10.4	(2.4)
Switzerland	10.2	(1.6)
Austria	10.0	(1.8)
Greece	9.8	(1.5)
Russian Federation	9.8	(2.7)
Scotland	9.0	(2.3)
Norway	8.9	(1.4)
New Zealand	8.7	(1.9)
Iceland	8.6	(2.2)
Canada	8.6	(1.4)
Singapore	8.2	(1.9)
Romania	8.2	(1.9)
Ireland	7.8	(2.1)
England	7.4	(1.7)
International	*7.3*	
Thailand	7.1	(2.3)
Australia	7.0	(1.7)
Spain	6.8	(1.3)
Latvia	6.6	(1.6)
Portugal	6.1	(1.2)
Germany	6.0	(1.8)
Netherlands	5.9	(2.5)
Japan	5.9	(1.0)
Iran	5.3	(1.3)
Hong Kong	5.1	(3.1)
Korea	4.0	(1.5)
USA	**3.9**	**(2.0)**
Belgium (Fr)	2.7	(1.9)
Colombia	2.0	(1.0)
Philippines	1.3	(1.3)
Bulgaria	0.8	(3.2)
South Africa	0.6	(1.4)
Belgium (Fl)	-0.1	(2.3)

Estimations & Errors

Country	Mean	(SE)
France	8.5	(1.3)
Spain	8.5	(1.5)
Scotland	8.3	(1.7)
New Zealand	7.6	(1.6)
Lithuania	7.5	(1.7)
England	7.4	(1.5)
Latvia	7.2	(1.7)
Denmark	7.2	(1.4)
Norway	6.9	(1.8)
Greece	6.9	(1.4)
Sweden	6.6	(1.2)
Switzerland	6.5	(1.3)
Ireland	6.5	(1.8)
Cyprus	6.4	(1.2)
Slovenia	5.8	(1.3)
Russian Federation	5.6	(1.5)
Canada	5.4	(1.1)
Australia	5.2	(1.3)
Portugal	5.2	(1.5)
Netherlands	4.9	(1.9)
International	*4.8*	
Bulgaria	4.7	(3.4)
USA	**4.7**	**(1.9)**
Hungary	4.7	(1.2)
Iran	4.3	(1.4)
Czech Republic	4.2	(1.4)
Hong Kong	4.0	(2.2)
Slovak Republic	3.9	(1.2)
Japan	3.9	(0.9)
Thailand	3.9	(1.6)
Singapore	3.5	(1.5)
Korea	3.0	(1.1)
Belgium (Fr)	2.9	(1.6)
Colombia	2.6	(2.0)
Iceland	2.6	(1.8)
Austria	2.1	(1.2)
Germany	1.7	(1.4)
Romania	1.4	(1.7)
Philippines	1.0	(1.5)
Belgium (Fl)	0.5	(1.4)
South Africa	0.1	(2.1)

Equations & Formulas

Country	Mean	(SE)
France	15.8	(1.3)
Spain	13.9	(1.1)
Slovenia	13.8	(1.2)
Norway	13.5	(0.9)
Switzerland	13.3	(1.0)
Greece	12.7	(1.1)
Canada	11.6	(1.0)
Slovak Republic	11.6	(1.3)
Austria	11.2	(1.2)
Netherlands	11.0	(2.1)
Denmark	10.9	(1.1)
Czech Republic	10.6	(2.0)
Hungary	10.6	(1.4)
Belgium (Fr)	9.9	(1.6)
Iceland	9.8	(1.4)
Germany	9.7	(2.0)
Scotland	9.7	(1.9)
New Zealand	9.6	(1.5)
Lithuania	9.6	(1.5)
Sweden	9.3	(1.0)
International	*8.9*	
Japan	8.9	(0.9)
Iran	8.8	(1.0)
Thailand	8.8	(2.2)
Latvia	8.5	(1.4)
Australia	8.4	(1.5)
Cyprus	8.3	(0.9)
Russian Federation	7.4	(1.8)
Singapore	7.3	(1.5)
Portugal	7.3	(1.1)
Ireland	7.2	(1.4)
England	7.0	(1.4)
USA	**6.9**	**(1.9)**
Korea	6.6	(1.0)
Romania	5.9	(1.6)
Hong Kong	5.3	(2.7)
Bulgaria	4.0	(2.4)
Colombia	3.5	(1.1)
South Africa	3.1	(1.4)
Philippines	2.7	(1.3)
Belgium (Fl)	2.3	(2.0)

Data Rep. & Analysis

Country	Mean	(SE)
Greece	11.7	(1.2)
Lithuania	9.3	(1.7)
Portugal	8.5	(1.1)
Latvia	8.3	(1.3)
Norway	7.5	(1.1)
Spain	7.2	(1.0)
New Zealand	7.1	(1.3)
Iran	6.8	(1.2)
Singapore	6.5	(1.3)
Czech Republic	6.4	(1.1)
Slovak Republic	6.2	(1.0)
Scotland	6.0	(1.6)
France	6.0	(1.1)
Denmark	6.0	(1.1)
Colombia	5.6	(1.7)
USA	**5.6**	**(1.5)**
Iceland	5.6	(1.3)
Ireland	5.4	(1.3)
Thailand	5.4	(1.3)
Bulgaria	5.3	(1.8)
Canada	5.3	(0.7)
Cyprus	5.3	(1.0)
International	*5.1*	
Sweden	4.9	(1.1)
Switzerland	4.9	(1.0)
Austria	4.7	(1.4)
Australia	4.7	(1.1)
Russian Federation	4.4	(1.5)
Slovenia	4.3	(0.9)
Romania	3.9	(1.4)
Japan	3.8	(0.6)
Hungary	3.7	(1.1)
England	3.3	(1.2)
Philippines	2.9	(1.2)
Germany	2.4	(1.5)
Korea	2.3	(0.8)
South Africa	2.1	(2.0)
Belgium (Fr)	2.0	(1.2)
Netherlands	1.9	(1.9)
Hong Kong	1.1	(1.8)
Belgium (Fl)	-0.9	(1.4)

Uncertainty & Probability

Country	Mean	(SE)
Iceland	9.7	(2.3)
New Zealand	9.4	(2.1)
France	8.9	(1.6)
Russian Federation	8.5	(2.0)
Scotland	8.4	(2.2)
Slovenia	8.4	(1.5)
Spain	8.4	(1.5)
Norway	8.3	(1.5)
Hungary	8.2	(1.6)
Greece	7.8	(1.6)
Lithuania	7.8	(1.8)
Singapore	7.6	(1.9)
Czech Republic	7.5	(2.1)
Switzerland	7.2	(1.4)
Latvia	7.0	(1.6)
Sweden	6.9	(1.7)
Germany	6.8	(2.0)
Bulgaria	6.8	(3.0)
Romania	6.5	(1.6)
Slovak Republic	6.4	(1.5)
Canada	6.4	(1.4)
Korea	6.4	(1.2)
Denmark	6.3	(1.8)
International	*6.3*	
Portugal	6.2	(1.3)
Australia	5.9	(1.7)
Austria	5.9	(1.5)
Iran	5.9	(1.7)
USA	**5.8**	**(2.1)**
Belgium (Fr)	5.7	(2.2)
Cyprus	5.6	(1.4)
Thailand	5.5	(1.9)
England	5.4	(1.9)
Japan	4.9	(1.0)
Colombia	4.8	(1.9)
Ireland	3.8	(1.9)
Hong Kong	3.5	(2.5)
Philippines	3.2	(1.4)
Netherlands	2.4	(2.4)
South Africa	0.5	(1.7)
Belgium (Fl)	0.2	(1.8)

Even using the overall achievement scores for mathematics and for science, we were at the bottom in terms of gains when represented by changes in country ranks on mean score. Perhaps this is what curricula that are 'a mile wide and an inch deep' typically produce. In any given content area, if new learning possibilities during a grade are only 'an inch deep', how could we expect anything other than small gains, even if these eventually reached a modest level of attainment cumulatively? We appear to have mathematics and science curricula in US education systems that were designed for small achievement gains in any one grade. If so, the TIMSS data suggest we got what our curricula were designed to produce.

In mathematics at the equivalent of eighth grade, almost half of the countries (18 of 41) ranked first or second in at least one of the 20 content areas discussed earlier. The US, as we said earlier, was among the seven countries that did not rank in the top 10 for any of the 20 areas. Without any technical subtleties to complicate the conclusion, this seems strong evidence to suggest that US eighth grade mathematics produced smaller gains than most countries in all content areas. Eighth grade mathematics curricula in US education systems 'worked' only if the comparatively small gains were part of cumulative attainment that eventually amounted to significantly high levels of achievement. We have already seen that this was not true in our previous discussions regarding achievement.

In eighth grade science, there were again only seven countries that were not among the top ten gainers for some content area examined. The US was one of those seven. Our best area was 'physical changes' where we ranked nineteenth among the 41 countries in achievement gains. Among all participating TIMSS countries, 13 out of 41 ranked either first or second for some area. Only four countries were not in the top 10 for either some area of mathematics or some area of science. These were the Philippines, South Africa, Germany, and the US. This strongly implies that there is an impact of curriculum on achievement gains.

At the fourth grade, the equivalent of 'the top ten' (roughly, the highest quartile) was the 'top six' countries since a smaller number of TIMSS countries participated in nine-year-old testing. In fourth grade mathematics, there were again seven countries that failed to make the top six in some content area. The US was one of those. Our best area was 'decimal fractions.' For this topic we ranked seventh out of the 24 countries who participated in testing nine-year-olds for both of the pair of appropriate grades. This was typically a topic first introduced in the US curriculum at this level so gains would hardly be surprising since we were most often going from 'not taught' to 'taught' for this topic.

In fourth grade science, there were only five countries that were not in the top six for some content areas. Finally, the US was *not* one of those five. However, the only area for which the US was among the top six in gains was

Exhibit 5.10. *Percentage Points Gain in Specific Science Areas for Eighth Grade Students (national mean percent gain).*

Earth Features

France	10.0	(1.1)
Cyprus	9.0	(0.9)
Thailand	8.5	(1.2)
Norway	8.3	(1.0)
Sweden	8.2	(1.0)
Lithuania	8.1	(1.3)
Greece	7.6	(1.0)
Scotland	7.0	(1.2)
Korea	6.9	(0.9)
Hong Kong	6.6	(1.5)
Austria	6.0	(1.0)
Romania	5.9	(1.7)
New Zealand	5.7	(1.1)
Switzerland	5.6	(0.9)
Netherlands	5.6	(1.5)
Latvia	5.5	(1.2)
Ireland	5.2	(1.2)
International	*5.2*	
Slovenia	5.1	(1.0)
Iran	4.9	(0.9)
Canada	4.9	(0.9)
Singapore	4.7	(1.8)
Czech Republic	4.6	(1.5)
Belgium (Fr)	4.6	(1.3)
Russian Federation	4.6	(1.3)
Australia	4.5	(1.2)
Philippines	4.2	(1.2)
Bulgaria	4.2	(1.8)
Hungary	4.2	(1.3)
Japan	4.2	(0.6)
Iceland	4.2	(1.3)
Colombia	4.0	(1.1)
England	3.8	(1.2)
USA	**3.7**	**(1.6)**
Denmark	3.6	(1.1)
Germany	3.5	(1.4)
Spain	3.4	(0.9)
Portugal	3.2	(1.0)
Slovak Republic	2.8	(1.2)
Belgium (Fl)	1.3	(1.3)
South Africa	-0.4	(1.3)

Earth Processes

Czech Republic	10.9	(1.9)
France	9.5	(1.3)
Slovak Republic	8.7	(1.3)
Sweden	8.3	(1.3)
Norway	8.3	(1.4)
Denmark	7.0	(1.5)
Austria	6.7	(1.7)
Belgium (Fr)	5.9	(1.6)
Thailand	5.9	(1.2)
Singapore	5.8	(1.9)
New Zealand	5.7	(1.5)
Portugal	5.3	(1.2)
Australia	5.3	(1.4)
England	5.2	(1.4)
Iceland	5.0	(2.0)
Greece	4.8	(1.0)
Spain	4.8	(1.1)
International	*4.8*	
Switzerland	4.7	(1.3)
Hungary	4.7	(1.3)
Romania	4.6	(1.7)
Ireland	4.5	(1.5)
Scotland	4.4	(1.5)
Lithuania	4.3	(1.5)
Slovenia	4.2	(1.1)
Japan	4.0	(0.9)
Germany	4.0	(1.7)
Colombia	3.8	(1.7)
Hong Kong	3.8	(1.7)
Bulgaria	3.6	(2.0)
Canada	3.1	(1.3)
Russian Federation	3.0	(1.3)
Cyprus	3.0	(1.3)
USA	**2.7**	**(1.7)**
Korea	2.7	(1.0)
Latvia	2.6	(1.4)
Belgium (Fl)	2.6	(2.2)
Netherlands	2.4	(2.1)
Philippines	2.2	(1.6)
Iran	2.2	(1.4)
South Africa	0.0	(1.8)

Earth in the Universe

Denmark	12.3	(2.3)
Hungary	9.7	(1.8)
Norway	9.2	(2.0)
Portugal	9.1	(1.9)
France	8.5	(1.7)
Iceland	8.2	(2.5)
Latvia	8.0	(2.0)
Lithuania	7.8	(2.5)
Slovenia	7.6	(1.7)
Sweden	7.3	(1.5)
Cyprus	7.3	(1.7)
Spain	7.1	(1.5)
Switzerland	6.5	(1.5)
Germany	6.5	(1.9)
New Zealand	6.5	(1.7)
Iran	6.1	(1.8)
Netherlands	4.9	(2.3)
International	*4.8*	
Australia	4.8	(1.5)
Greece	4.4	(1.6)
Japan	4.4	(1.2)
Romania	4.4	(1.6)
Colombia	4.3	(2.0)
Belgium (Fr)	4.1	(2.2)
Ireland	4.0	(1.6)
Austria	3.4	(1.8)
Scotland	3.2	(1.9)
Hong Kong	3.1	(2.3)
USA	**3.1**	**(1.9)**
Russian Federation	2.9	(1.5)
Canada	2.8	(1.5)
Singapore	2.6	(1.5)
Czech Republic	2.4	(1.9)
Thailand	2.4	(2.1)
Slovak Republic	1.9	(1.7)
Bulgaria	1.8	(3.7)
Philippines	0.9	(1.8)
England	0.3	(2.0)
Korea	0.0	(1.2)
South Africa	-0.1	(2.1)
Belgium (Fl)	-0.5	(2.0)

Structure of Matter

Lithuania	31.5	(1.8)
Singapore	24.2	(2.0)
Russian Federation	19.5	(2.4)
Greece	18.4	(1.3)
Sweden	18.2	(1.5)
Portugal	17.2	(1.4)
Latvia	15.3	(1.8)
Bulgaria	13.3	(3.0)
Spain	12.5	(1.6)
Australia	11.6	(1.5)
Norway	11.1	(1.4)
Thailand	10.5	(1.5)
Iran	10.5	(1.6)
Ireland	10.1	(1.8)
Netherlands	9.6	(1.9)
International	*9.1*	
New Zealand	9.0	(1.6)
Germany	8.9	(1.9)
Japan	8.6	(1.3)
England	7.9	(1.9)
USA	**7.9**	**(1.6)**
Romania	7.9	(2.0)
Slovak Republic	7.5	(1.8)
Switzerland	7.3	(1.2)
France	6.7	(1.5)
Korea	6.6	(1.3)
Czech Republic	6.3	(2.4)
Cyprus	6.1	(1.3)
Denmark	5.9	(1.5)
Austria	5.5	(2.0)
Scotland	4.6	(1.8)
Canada	4.2	(1.3)
Philippines	4.1	(1.3)
Hungary	3.9	(1.7)
Iceland	3.8	(1.6)
Belgium (Fr)	3.6	(1.6)
Colombia	2.6	(2.2)
Hong Kong	1.5	(1.6)
South Africa	0.9	(1.2)
Belgium (Fl)	0.4	(1.4)
Slovenia	-0.4	(1.7)

Energy & Physical Processes

Russian Federation	8.9	(1.2)
Latvia	8.6	(0.9)
Lithuania	8.4	(0.9)
Portugal	8.2	(0.7)
Greece	8.0	(0.7)
Netherlands	7.2	(1.4)
Cyprus	7.0	(0.6)
Singapore	6.8	(1.3)
Czech Republic	6.8	(1.1)
Denmark	6.7	(1.0)
Norway	6.6	(0.8)
Iran	6.6	(1.0)
New Zealand	6.6	(1.0)
Scotland	6.5	(1.1)
Slovak Republic	6.3	(0.9)
Austria	6.1	(1.1)
Ireland	6.1	(1.1)
Switzerland	5.9	(0.7)
France	5.8	(0.8)
Slovenia	5.8	(0.8)
Belgium (Fr)	5.5	(0.9)
International	*5.4*	
Hungary	5.3	(0.8)
Australia	5.3	(1.0)
Spain	5.1	(0.7)
Sweden	4.9	(0.8)
Romania	4.8	(1.1)
Canada	4.6	(0.7)
Iceland	4.3	(1.2)
England	4.3	(1.0)
Bulgaria	4.3	(1.5)
Japan	3.7	(0.5)
Hong Kong	3.6	(1.5)
Thailand	3.6	(1.0)
USA	**3.6**	**(1.3)**
Germany	3.4	(1.2)
Belgium (Fl)	3.2	(1.2)
Colombia	2.4	(1.2)
Korea	2.2	(0.7)
Philippines	1.8	(1.2)
South Africa	1.7	(1.7)

Physical Changes

France	14.8	(2.4)
Lithuania	14.3	(2.7)
Portugal	13.6	(1.8)
Spain	12.5	(2.0)
New Zealand	11.3	(2.1)
Greece	10.3	(1.6)
Russian Federation	10.2	(2.7)
Norway	9.9	(2.7)
Hungary	9.7	(2.0)
Iceland	9.3	(3.4)
Iran	9.0	(2.5)
Sweden	8.9	(1.9)
Latvia	8.9	(2.3)
Austria	8.6	(2.4)
Denmark	8.4	(2.4)
Cyprus	8.3	(2.4)
Netherlands	7.9	(2.8)
Slovak Republic	7.8	(2.0)
Slovenia	6.8	(2.2)
International	*6.2*	
USA	**5.9**	**(2.4)**
Ireland	5.2	(2.4)
Czech Republic	5.1	(2.5)
Canada	5.1	(2.3)
Germany	4.7	(2.5)
Philippines	4.5	(2.0)
Thailand	4.5	(1.9)
Belgium (Fr)	4.3	(2.4)
Romania	4.1	(2.0)
Colombia	3.6	(2.3)
Switzerland	3.2	(1.8)
Australia	2.5	(1.9)
Japan	2.0	(1.7)
Belgium (Fl)	1.6	(2.4)
South Africa	1.6	(2.4)
Scotland	1.3	(2.2)
England	1.1	(2.7)
Hong Kong	0.7	(1.3)
Bulgaria	0.1	(4.8)
Singapore	-0.4	(2.1)
Korea	-3.1	(1.8)

Exhibit 5.10. (Continued)

Diversity & Structure of Living Things		
Lithuania	11.7	(1.4)
Japan	9.8	(0.6)
Russian Federation	9.5	(1.4)
Portugal	8.8	(1.0)
Sweden	8.1	(1.1)
Singapore	7.8	(1.7)
Latvia	7.6	(1.3)
Scotland	6.9	(1.4)
Belgium (Fr)	6.6	(1.3)
Ireland	6.5	(1.4)
New Zealand	6.4	(1.3)
Hungary	6.3	(1.1)
Cyprus	6.1	(1.0)
Iceland	6.1	(1.5)
France	6.1	(1.1)
Australia	5.9	(1.2)
Denmark	5.5	(1.2)
England	5.4	(1.1)
Norway	5.3	(1.3)
International	*5.2*	
Spain	5.2	(0.8)
Austria	5.1	(1.2)
Germany	5.1	(1.5)
Greece	5.1	(1.0)
Switzerland	5.0	(1.1)
Bulgaria	5.0	(2.0)
Czech Republic	5.0	(1.2)
Netherlands	4.5	(1.6)
Korea	4.4	(1.0)
Slovak Republic	4.3	(1.0)
Colombia	3.1	(1.5)
Canada	3.0	(0.9)
Slovenia	3.0	(1.1)
Romania	2.7	(1.4)
Hong Kong	2.6	(1.5)
Thailand	2.5	(1.2)
Iran	2.4	(1.1)
USA	2.0	(1.5)
Philippines	1.7	(1.4)
Belgium (Fl)	1.2	(1.5)
South Africa	0.3	(1.7)

Life Processes & Functions		
Lithuania	13.5	(1.6)
Singapore	10.7	(1.9)
Russian Federation	9.9	(1.5)
Latvia	9.5	(1.3)
Iceland	9.0	(1.5)
New Zealand	8.7	(1.5)
Norway	8.5	(1.2)
Japan	8.5	(0.8)
Scotland	8.4	(1.6)
Ireland	8.2	(1.5)
Denmark	8.0	(1.4)
Australia	7.9	(1.3)
Hong Kong	7.8	(1.9)
Sweden	7.8	(1.2)
Belgium (Fr)	7.7	(1.5)
Switzerland	7.3	(1.0)
Portugal	7.1	(1.2)
Spain	7.1	(1.0)
England	6.9	(1.3)
France	6.9	(1.5)
Greece	6.8	(1.2)
International	*6.7*	
Belgium (Fl)	6.5	(1.7)
Cyprus	5.9	(1.1)
Colombia	5.8	(1.5)
Canada	5.8	(1.0)
Austria	5.7	(1.5)
Korea	5.6	(1.0)
Netherlands	5.6	(2.1)
Czech Republic	5.6	(1.5)
Slovak Republic	5.6	(1.2)
Slovenia	5.4	(1.2)
Thailand	5.2	(1.5)
Germany	4.9	(1.8)
Bulgaria	4.8	(2.2)
Romania	4.5	(1.5)
USA	4.0	(1.6)
Iran	3.4	(1.3)
Hungary	3.2	(1.2)
Philippines	1.4	(1.5)
South Africa	1.3	(1.9)

Life Cycles & Genetics		
Singapore	10.3	(2.1)
Lithuania	10.1	(2.0)
France	8.0	(1.3)
Latvia	7.9	(1.7)
Belgium (Fr)	7.9	(1.8)
England	7.5	(2.2)
Ireland	7.4	(1.8)
Russian Federation	7.3	(1.5)
Cyprus	7.1	(1.7)
Iceland	7.1	(2.1)
Scotland	7.1	(1.8)
Hungary	7.0	(1.5)
Australia	6.9	(1.5)
Sweden	6.9	(1.3)
Thailand	6.1	(1.6)
Slovenia	6.0	(1.6)
New Zealand	5.9	(1.7)
Denmark	5.8	(1.6)
Spain	5.6	(1.4)
USA	4.9	(1.6)
International	*4.8*	
Romania	4.8	(2.0)
Netherlands	4.7	(2.4)
Portugal	4.7	(1.4)
Czech Republic	4.3	(1.5)
Germany	3.7	(1.8)
Canada	3.6	(1.1)
Bulgaria	3.4	(2.8)
Greece	3.3	(1.5)
Norway	3.3	(1.5)
Korea	3.0	(1.3)
Switzerland	2.9	(1.4)
Slovak Republic	2.8	(1.2)
Hong Kong	2.8	(1.9)
Austria	2.1	(1.7)
Colombia	1.9	(2.1)
Iran	1.4	(2.1)
Belgium (Fl)	0.4	(1.8)
Japan	0.1	(1.0)
Philippines	-0.4	(1.4)
South Africa	-2.3	(1.8)

Chemical Changes		
Lithuania	19.0	(1.5)
Russian Federation	17.2	(2.0)
Portugal	16.4	(1.3)
Japan	13.4	(0.7)
Latvia	13.1	(1.4)
Scotland	11.4	(1.7)
New Zealand	10.6	(1.3)
Korea	10.4	(1.1)
Slovak Republic	9.7	(1.3)
Cyprus	9.3	(1.1)
Spain	9.2	(1.1)
Sweden	8.8	(1.0)
Austria	8.8	(1.5)
Czech Republic	8.6	(1.5)
Greece	8.5	(1.0)
Netherlands	8.5	(1.3)
Belgium (Fl)	8.1	(1.3)
International	*7.9*	
Hungary	7.8	(1.1)
Australia	7.7	(1.2)
Germany	7.7	(1.6)
Switzerland	7.4	(1.1)
England	7.1	(1.5)
Iran	6.7	(1.4)
France	6.5	(1.4)
Ireland	6.4	(1.5)
Canada	6.1	(1.1)
Norway	5.9	(1.2)
Iceland	5.5	(1.4)
Thailand	5.4	(1.5)
Denmark	5.4	(1.1)
Singapore	5.4	(1.8)
Bulgaria	5.2	(1.9)
Hong Kong	5.2	(1.8)
USA	5.0	(1.6)
Romania	4.3	(1.5)
Philippines	4.1	(1.4)
South Africa	3.9	(1.8)
Slovenia	3.8	(1.4)
Belgium (Fr)	3.0	(1.2)
Colombia	0.6	(1.6)

Forces & Motion		
Lithuania	13.6	(2.2)
Netherlands	11.9	(2.0)
Greece	10.0	(1.6)
Sweden	9.4	(1.6)
Latvia	9.2	(2.0)
Scotland	8.8	(2.2)
Portugal	8.6	(1.6)
Iran	8.5	(2.2)
Spain	8.3	(1.5)
Cyprus	8.0	(1.6)
Switzerland	7.1	(1.6)
Colombia	6.8	(2.2)
France	6.5	(1.8)
Singapore	6.5	(1.5)
New Zealand	6.4	(1.6)
England	6.0	(2.0)
Australia	5.9	(1.6)
Germany	5.9	(2.1)
Slovenia	5.8	(1.7)
International	*5.6*	
Ireland	5.6	(1.9)
Norway	5.5	(2.1)
Hungary	5.5	(1.8)
Austria	5.4	(1.8)
Philippines	5.3	(1.6)
USA	5.0	(2.1)
Korea	4.8	(1.4)
Hong Kong	4.5	(2.1)
Romania	3.9	(1.9)
Canada	3.8	(1.7)
Belgium (Fr)	3.6	(2.1)
Japan	3.3	(1.2)
Bulgaria	3.1	(2.9)
Denmark	2.8	(1.9)
Thailand	2.7	(1.7)
Slovak Republic	2.5	(1.6)
South Africa	2.1	(2.0)
Czech Republic	2.0	(1.5)
Russian Federation	1.7	(1.8)
Iceland	0.9	(3.0)
Belgium (Fl)	(2.7)	(3.0)

Science, Technology, & Society		
France	13.5	(2.5)
Greece	11.4	(2.1)
Iceland	11.1	(3.6)
Singapore	11.0	(2.8)
Norway	10.4	(2.4)
Colombia	10.2	(2.5)
Belgium (Fr)	9.9	(3.0)
Iran	9.9	(2.7)
Austria	9.8	(2.8)
Denmark	8.6	(2.7)
Scotland	8.4	(2.5)
Japan	8.2	(1.7)
Hong Kong	7.6	(3.3)
Switzerland	7.2	(2.3)
New Zealand	6.5	(2.5)
Netherlands	6.4	(3.3)
Portugal	6.3	(2.0)
International	*6.1*	
Australia	6.1	(2.1)
Ireland	6.0	(2.4)
Thailand	5.9	(2.4)
Spain	5.9	(2.3)
Korea	5.9	(2.6)
Canada	5.7	(2.6)
Latvia	5.5	(2.3)
Germany	5.3	(3.3)
Lithuania	5.3	(3.0)
Sweden	4.9	(2.4)
Romania	4.2	(2.1)
England	4.1	(2.6)
Bulgaria	4.0	(4.5)
Slovenia	3.9	(2.7)
Russian Federation	3.5	(2.8)
Slovak Republic	2.9	(2.3)
Belgium (Fl)	2.5	(2.9)
Cyprus	1.8	(2.1)
USA	1.7	(2.2)
Philippines	1.5	(2.4)
South Africa	0.9	(2.0)
Hungary	0.9	(3.8)
Czech Republic	0.8	(2.7)

Exhibit 5.10. *(Continued)*

Interactions of Living Things		
Latvia	11.9	(2.0)
Romania	10.5	(2.1)
Norway	10.3	(2.2)
Lithuania	10.2	(2.1)
Russian Federation	9.5	(2.2)
Switzerland	9.2	(1.6)
Iceland	9.2	(3.0)
England	9.1	(2.2)
Singapore	8.7	(2.1)
New Zealand	8.2	(2.1)
Greece	8.0	(1.8)
Denmark	7.8	(2.0)
Ireland	7.8	(2.1)
Canada	7.8	(1.7)
Australia	7.7	(1.9)
Portugal	7.5	(1.7)
Slovak Republic	7.4	(1.8)
Belgium (Fl)	7.2	(2.2)
Czech Republic	7.1	(2.0)
Scotland	6.9	(2.4)
Colombia	6.8	(3.1)
Netherlands	6.7	(2.1)
Austria	6.6	(2.1)
Spain	6.5	(1.7)
Hong Kong	6.4	(2.2)
International	6.2	
Iran	5.5	(1.8)
Cyprus	5.5	(1.7)
Germany	4.7	(2.2)
Slovenia	4.7	(2.1)
USA	4.6	(2.1)
Japan	4.6	(1.1)
France	4.2	(1.9)
Korea	4.2	(1.7)
Belgium (Fr)	4.1	(2.0)
Hungary	4.1	(2.0)
Sweden	3.9	(1.7)
Thailand	3.5	(1.9)
South Africa	0.7	(2.1)
Philippines	-1.1	(1.7)
Bulgaria	-8.9	(3.5)

Human Biology & Health		
Singapore	12.5	(1.8)
Lithuania	11.4	(1.3)
Portugal	10.8	(0.9)
Ireland	9.8	(1.4)
Cyprus	9.3	(1.0)
Scotland	8.6	(1.5)
Japan	8.3	(0.7)
Switzerland	8.0	(0.7)
Russian Federation	7.8	(1.2)
New Zealand	7.4	(1.2)
Australia	7.0	(1.1)
Latvia	6.7	(1.1)
Netherlands	6.7	(1.8)
France	6.6	(1.0)
Korea	6.6	(0.9)
Denmark	6.5	(1.1)
Austria	6.3	(1.1)
Bulgaria	6.3	(1.5)
England	6.2	(1.1)
Germany	6.2	(1.4)
Thailand	5.9	(1.2)
Greece	5.8	(0.9)
International	5.8	
Sweden	5.4	(1.0)
Canada	5.2	(0.9)
Spain	5.2	(0.8)
Colombia	5.1	(1.8)
Belgium (Fr)	5.1	(1.3)
Hungary	4.4	(1.0)
Hong Kong	4.0	(1.5)
USA	3.9	(1.5)
Iceland	3.9	(1.2)
Czech Republic	3.6	(1.2)
Iran	3.6	(1.1)
Norway	3.4	(1.0)
Slovak Republic	2.9	(0.9)
Belgium (Fl)	2.2	(1.4)
Slovenia	2.0	(1.1)
Romania	1.6	(1.3)
South Africa	0.2	(1.9)
Philippines	-1.3	(1.5)

Properties & Classification of Matter		
Portugal	14.0	(1.1)
Lithuania	13.4	(1.4)
Latvia	10.7	(1.2)
France	10.5	(1.3)
Russian Federation	9.2	(1.3)
New Zealand	8.7	(1.3)
Greece	8.4	(1.0)
Netherlands	8.1	(1.7)
Sweden	8.0	(1.1)
Denmark	7.5	(1.3)
Iceland	7.2	(1.8)
Scotland	7.0	(1.5)
Norway	7.0	(1.7)
Austria	6.8	(1.5)
Canada	6.3	(1.1)
Spain	6.1	(1.1)
Germany	6.1	(1.6)
Switzerland	6.1	(1.1)
Slovak Republic	6.0	(1.2)
International	5.9	
Hong Kong	5.6	(1.7)
Iran	5.5	(1.4)
Ireland	5.1	(1.4)
Belgium (Fr)	4.9	(1.3)
Slovenia	4.6	(1.2)
Romania	4.4	(1.2)
Belgium (Fl)	4.2	(2.0)
Hungary	4.2	(1.3)
Singapore	4.1	(1.8)
Japan	3.9	(0.8)
USA	3.9	(1.5)
Cyprus	3.6	(1.1)
Colombia	3.5	(1.5)
Bulgaria	3.5	(2.0)
Philippines	3.1	(1.5)
Czech Republic	2.6	(1.4)
Australia	2.6	(1.2)
Thailand	2.5	(1.5)
South Africa	2.3	(1.8)
England	1.9	(1.3)
Korea	1.1	(0.9)

Environmental & Resource Issues		
Singapore	11.9	(1.5)
France	8.0	(1.5)
Austria	7.6	(1.6)
Scotland	7.6	(1.8)
Japan	7.4	(1.0)
Greece	7.4	(1.2)
Denmark	6.9	(1.4)
Lithuania	6.6	(1.7)
Norway	6.4	(1.5)
England	6.2	(1.8)
Czech Republic	6.1	(1.9)
Spain	6.1	(1.3)
Hong Kong	6.0	(1.9)
Colombia	6.0	(1.6)
Bulgaria	5.9	(2.2)
Netherlands	5.8	(2.3)
Portugal	5.7	(1.2)
Australia	5.6	(1.4)
Canada	5.5	(1.2)
Latvia	5.5	(1.6)
Belgium (Fr)	5.3	(1.5)
Sweden	5.3	(1.4)
Ireland	5.3	(1.7)
Romania	5.2	(1.6)
International	5.2	
Russian Federation	5.0	(1.5)
Switzerland	5.0	(1.3)
Iceland	4.8	(1.9)
Thailand	4.6	(1.5)
USA	4.2	(1.8)
New Zealand	3.8	(1.5)
Belgium (Fl)	3.7	(1.5)
Germany	3.6	(1.9)
Cyprus	3.6	(1.4)
Hungary	3.4	(1.6)
Slovenia	2.6	(1.6)
Philippines	2.3	(1.4)
Korea	2.3	(1.4)
Iran	2.3	(2.0)
Slovak Republic	1.7	(1.5)
South Africa	0.2	(1.9)

Scientific Processes		
Latvia	12.4	(1.9)
Singapore	11.1	(2.3)
Portugal	10.2	(1.4)
Cyprus	9.6	(1.6)
Scotland	9.1	(2.2)
England	9.0	(1.8)
France	9.0	(1.4)
Lithuania	8.9	(1.6)
Bulgaria	8.9	(3.0)
Russian Federation	8.1	(1.8)
Greece	7.8	(1.6)
Iceland	7.8	(1.7)
Norway	7.5	(1.7)
Denmark	7.5	(2.1)
Iran	7.2	(1.4)
Netherlands	7.0	(2.5)
Sweden	6.8	(1.3)
Ireland	6.7	(1.6)
New Zealand	6.6	(2.0)
Germany	6.5	(1.8)
International	6.3	
Spain	6.3	(1.3)
Belgium (Fr)	6.0	(1.9)
Austria	6.0	(1.7)
USA	5.8	(1.9)
Slovak Republic	5.7	(1.4)
Canada	5.6	(1.3)
Czech Republic	5.2	(1.9)
Australia	5.2	(1.6)
Hungary	5.0	(1.5)
Switzerland	4.8	(1.4)
Romania	4.4	(1.6)
Belgium (Fl)	4.2	(2.2)
Japan	4.1	(1.2)
Thailand	3.9	(1.9)
Slovenia	3.4	(1.5)
Colombia	3.1	(1.8)
Korea	2.8	(1.3)
Hong Kong	2.7	(2.4)
Philippines	1.1	(1.4)
South Africa	1.0	(1.8)

for science 'process' items. The highest content area was 'organs and tissues' for which we achieved a rank of eleventh out of 24 rankable countries. Close to half of the counties, 10 of 24 countries testing at both grades, ranked first or second for some content area.

Again for fourth grade, there were few countries that were not in the top six in some content area in either mathematics or science. Three countries attained this dubious distinction: England, Scotland, and Thailand. The US would join these three countries if we focused only on content areas strictly defined, and excluded the 'content free' process-focused items that did not draw on factual science knowledge.

One could argue that an effective curriculum should produce major gains in at least some area at each grade. Even if students did well in many areas because of cumulative gains over several years, it seems likely that there would be major gains at some grade level for at least one content area. This would seem to be particularly true if the curriculum had different content areas that received focused instructional attention at different grade levels. This appears to be the case for almost all of the TIMSS countries.

Only a curriculum that was composed largely without such focused attention would consistently produce at best modest gains across the content areas. However, this is exactly the pattern that is true for the US in both science and mathematics at fourth and eighth grades. This distinction is reinforced by the fact that of the 26 countries who tested both nine- and thirteen-year-olds, the US was the only country not a part of the top gaining countries in science or mathematics at either fourth or eighth grade for any content area (excluding the 'scientific processes' area).

The US also was distinguished by the consistency of its mathematics and science curricula at all levels in being 'a mile wide and an inch deep', rarely allocating focused attention to specific topics. We have adopted a unique curricular strategy – or perhaps 'strategies' since the means of being a 'mile wide and an inch deep' differ for mathematics and science. Given the concomitant achievement gains linked with that curricular strategy at fourth and eighth grades, we can only consider that approach risky and unproductive since it was associated with only modest gains and average cumulative achievements in most content areas and in situations when curricular differences were the only likely explanation for achievement differences.

While cause and effect can not be unambiguously assigned in this kind of study, the US achievement data and curriculum data are at least consistent with this possibility. This poses a pressing question, 'Do these small gains eventually accumulate to produce the overall gains comparable to those in countries that focus on specific content areas at specific grade levels?' The answer through eighth grade is 'no' for both science and mathematics.

Item Level Gains. Do the gains seen for specific items correspond to those seen for content areas? We can again use median polish techniques to establish typical gains on each item across countries and for each country across items. From this we can identify the items for which any given country had gains that were higher or lower than would be expected from the general level of gains for that country and item.

In fourth grade mathematics over all TIMSS countries, about two-thirds of the items had unusually high or low gains which indicates a consistently high level of variability in gains for most items. Individual items had quite distinct patterns of gains in different countries. Again, for the US, 'data representation' items contained several of those in which we did better than expected. This is consistent with the fact that it was the content area in which we achieved our greatest gains, ranking ninth among the 24 rankable countries (of 26 which gave the test, two did not provide sufficient data for gain computations).

In science, there was only one out of more than 150 items on which our gain was exceptional compared to our typical gains over all items and each item's typical gains across countries. This was a relatively science-knowledge-free 'process' item. Nothing in the item level gain results was inconsistent with the content area results previously discussed.

Country by item interactions are another source of evidence that curricular differences had effects on achievement gains. In both fourth and eighth grades, and for both science and mathematics, there were many such interactions. This would seem to imply significant curricular effects, compellingly so since these could be linked to a single year's schooling.

If these interactions are one indicator of curricular effects, the absence of such effects would seem to imply that children in a country gain at the same level on all items, once we account for item difficulties. This would suggest that there was probably not much difference in coverage across topic areas – in the access to educational possibilities for that country as a whole.

However, in fourth grade mathematics most countries had between seven and 20 items with unexpected gain measures. This suggests some form of differentiated learning opportunities or coverage across topic areas. The US had only three. Other countries with a similar small number of interactions included Australia, Canada, England, New Zealand, and Scotland. In science, the number of items with unexpected gains was typically from five to 15. The US had only one such item.

These comparative results could be taken as additional evidence supporting the earlier suggestion that this uniformity for items (and concomitant modest gains) was the consequence of mathematics and science curricula that were 'a mile wide and an inch deep.' If virtually all areas receive similarly brief attention in a given grade, then we would expect similar small gains for all but anomalous achievement items.

Additionally, we could suggest the hypothesis that curricula with such even attention across a broad range of topics and consistent small gains across all areas and items actually fosters a situation in which factors unrelated to curriculum could dominate curricular effects on test scores, especially total test scores – factors such as student motivation, social class, and so on. This would imply that the nature of the US curriculum reinforced well established global linkages between achievement and socio-economic status.

Gains on items linking all four tested grades. There were a few specific items that were the same for students tested at the equivalent of third, fourth, seventh, and eighth grades. Here we present several items in mathematics and in science.

One mathematics item (shown in Exhibit 5.11) dealt with measurement and showed only modest change in the percentage of students who got it right (between 25 and just over 30 percent) from third grade to eighth grade. This contrasted with a change in the cross-national means that doubled (from about 21 to about 42 percent correct) over the same span. This was in a content area

Exhibit 5.11. Results For A Measurement Item Appearing on Tests for Third, Fourth, Seventh, and Eighth Grade Students.

	Grade 3		Grade 4		Grade 7		Grade 8	
USA	25.1	(2.7)	23.3	(1.9)	29.3	(2.1)	31.5	(1.3)
International	21.7		23.8		37.7		43.2	
Australia	24.2	(2.6)	23.3	(2.2)	42.3	(1.7)	49.9	(1.4)
Canada	19.3	(1.9)	22.8	(2.4)	35.0	(1.5)	44.3	(1.6)
Czech Republic	15.3	(1.8)	16.1	(1.8)	42.2	(2.0)	49.3	(2.1)
England	21.2	(2.6)	29.0	(3.2)	41.7	(1.6)	48.4	(1.9)
Hong Kong	20.1	(1.8)	28.9	(1.9)	53.9	(2.7)	58.8	(2.4)
Hungary	13.4	(1.8)	15.2	(2.0)	25.4	(1.5)	33.2	(1.8)
Israel			17.1	(2.5)			44.6	(3.0)
Japan	33.3	(2.2)	32.2	(2.2)	59.8	(1.2)	69.8	(1.1)
Korea	36.7	(2.6)	38.3	(3.0)	56.6	(1.9)	63.5	(1.4)
Netherlands	31.0	(2.8)	35.3	(3.6)	51.5	(2.2)	48.4	(2.3)
New Zealand	25.2	(2.4)	22.7	(2.3)	39.9	(1.9)	45.1	(1.8)
Norway	18.4	(2.7)	18.5	(2.2)	30.0	(2.1)	36.3	(1.6)
Singapore	18.5	(1.4)	46.5	(2.0)	74.0	(2.5)	83.3	(1.4)
Thailand	12.4	(2.1)	14.7	(2.4)	30.6	(1.9)	37.5	(2.4)

E6.   A thin wire 20 centimeters long is formed into a rectangle. If the width of this rectangle is 4 centimeters, what is its length?

A.   5 centimeters

B.   6 centimeters

C.   12 centimeters

D.   16 centimeters

in which US students consistently did not do well. US third graders were about at the international average, but there was a smaller difference between the third and fourth grade averages for the US than internationally.

A second item (see Exhibit 5.12) showed gains of about 35 percent more US children who correctly answered the item at eighth grade than at third grade. US achievement was higher than the cross-national mean at all grades. The growth for the content area that contained this item was more modest (as would be expected for a cluster of items rather than a single item) and US achievement and gains were more comparable.

Exhibit 5.13 presents a science item related to the physical science topic of magnetism. US achievement at each grade level and overall gains are comparable to the cross-national means. Here, there is a moderate increase between US third and fourth grades and a comparable gain between fourth and seventh. These gains on each of the three items (Exhibit 5.11, Exhibit 5.12,

Exhibit 5.12. *Results for an Item Combining Estimation and Number Sense Appearing on Tests for Third, Fourth, Seventh, and Eighth Grade Students.*

	Grade 3		Grade 4		Grade 7		Grade 8	
USA	42.8	(1.7)	63.5	(1.5)	77.3	(1.9)	76.7	(1.8)
International	30.9		48.8		66.1		69.6	
Australia	32.2	(1.4)	48.1	(1.3)	73.5	(2.5)	76.9	(1.7)
Canada	29.7	(1.4)	52.7	(1.7)	81.3	(1.8)	82.5	(1.7)
Czech Republic	36.3	(1.9)	59.2	(1.7)	83.4	(2.0)	86.2	(2.1)
England	28.9	(1.3)	36.0	(1.6)	63.4	(3.6)	73.1	(2.9)
Hong Kong	37.0	(1.7)	72.5	(1.3)	80.7	(2.6)	82.1	(2.3)
Hungary	35.5	(1.7)	58.9	(1.7)	71.5	(2.3)	79.8	(2.6)
Israel			58.0	(2.4)			75.2	(3.2)
Japan	28.2	(1.1)	64.6	(1.2)	83.0	(1.6)	83.6	(1.6)
Korea	35.9	(1.2)	68.0	(1.3)	81.4	(2.1)	80.3	(2.2)
Netherlands	33.7	(1.4)	58.0	(1.9)	77.1	(2.2)	81.0	(3.0)
New Zealand	27.2	(1.4)	36.6	(1.5)	57.9	(3.0)	72.8	(2.1)
Norway	23.3	(1.4)	34.6	(1.5)	56.3	(4.5)	65.1	(2.6)
Singapore	38.6	(1.0)	76.2	(1.3)	91.0	(1.6)	91.5	(1.4)
Thailand	33.6	(1.2)	32.7	(1.4)	70.3	(2.6)	69.9	(2.1)

P12. Mark's garden has 84 rows of cabbages. There are 57 cabbages in each row. Which of these gives the BEST way to estimate how many cabbages there are altogether?

A. $100 \times 50 = 5,000$

B. $90 \times 60 = 5,400$

C. $80 \times 60 = 4,800$

D. $80 \times 50 = 4,000$

Exhibit 5.13. *Results for a Physical Science Item Appearing on Tests for Third, Fourth, Seventh, and Eighth Grade Students.*

	Grade 3		Grade 4		Grade 7		Grade 8	
USA	38.5	(2.9)	53.0	(2.3)	73.1	(1.9)	74.5	(1.2)
International	*41.9*		*50.7*		*68.5*		*75.8*	
Australia	43.5	(3.0)	56.3	(2.5)	78.2	(1.6)	81.4	(1.0)
Canada	33.1	(2.4)	50.6	(2.5)	72.5	(1.3)	79.4	(1.0)
Czech Republic	42.3	(2.7)	65.8	(2.8)	90.3	(0.9)	94.1	(1.1)
Hong Kong	46.9	(2.5)	58.8	(1.9)	73.8	(1.7)	77.7	(1.4)
Hungary	55.6	(2.7)	64.1	(3.3)	87.2	(1.2)	93.7	(0.8)
Israel			38.3	(2.9)			68.5	(2.8)
Japan	74.9	(1.8)	82.8	(1.6)	84.1	(1.0)	88.9	(0.9)
Korea	73.3	(2.3)	83.6	(2.2)	86.5	(1.0)	86.2	(1.0)
Netherlands	32.0	(2.8)	52.8	(3.1)	71.1	(2.1)	82.0	(1.4)
New Zealand	41.0	(3.0)	52.5	(3.1)	61.8	(1.7)	74.2	(1.5)
Norway	36.5	(3.2)	40.2	(3.1)	60.8	(2.5)	73.5	(1.4)
Singapore	56.6	(2.0)	70.3	(1.8)	87.4	(1.3)	90.8	(0.8)
Thailand	26.3	(3.3)	29.4	(3.6)	64.1	(1.5)	71.3	(1.6)
England	48.7	(2.9)	59.5	(1.9)	81.1	(1.9)	88.2	(1.4)

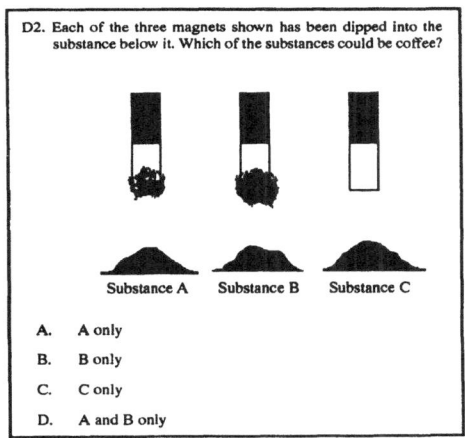

D2. Each of the three magnets shown has been dipped into the substance below it. Which of the substances could be coffee?

Substance A   Substance B   Substance C

A. A only

B. B only

C. C only

D. A and B only

and Exhibit 5.13) could be due to schooling and curricular effects, to an accumulation of life experience over several years, to developmental differences, or to some combination of these factors. Disentangling these different effects is beyond the scope of the analyses presented here. The fact that the gains have various magnitudes for the grade pairs in different countries is not inconsistent with curricular effects.

Exhibit 5.14 presents a second physical science item with a somewhat different pattern of gains. In all cases, US students' performance was below the cross-national means. Here there is a modest gain between third and fourth grades, and even a small gain from seventh and eighth grade. There is a much larger gain between fourth and seventh.

Exhibit 5.14. *Results For Another Physical Science Item Appearing on Tests for Third, Fourth, Seventh, and Eighth Grade Students.*

	Grade 3	Grade 4	Grade 7	Grade 8
USA	11.9 (1.6)	20.7 (1.8)	38.5 (3.1)	41.6 (2.2)
International	*16.1*	*21.9*	*47.4*	*52.8*
Australia	15.3 (1.8)	19.6 (1.6)	47.8 (2.3)	57.2 (2.0)
Canada	13.0 (1.8)	22.3 (1.8)	50.9 (2.9)	51.8 (2.2)
Czech Republic	21.7 (2.5)	28.2 (2.6)	76.6 (2.3)	74.1 (2.7)
Hong Kong	18.7 (1.9)	28.4 (2.6)	62.2 (2.6)	70.5 (2.8)
Hungary	16.4 (1.8)	26.3 (2.7)	54.4 (3.0)	62.0 (2.9)
Israel		13.1 (2.2)		35.7 (3.8)
Japan	18.5 2.0	27.4 (2.1)	53.7 2.0	60.8 (1.7)
Korea	22.4 (2.6)	25.9 (2.7)	42.6 (2.9)	52.4 (3.0)
Netherlands	14.9 (2.4)	27.9 (2.4)	47.2 (3.0)	65.3 (3.3)
New Zealand	12.7 (2.3)	17.2 (2.5)	47.5 (2.9)	55.2 (2.5)
Norway	17.4 (2.1)	21.1 (3.0)	50.7 (3.3)	60.5 (2.4)
Singapore	15.4 (1.2)	32.0 (1.7)	65.4 (2.5)	75.8 (2.2)
Thailand	7.6 (1.6)	15.4 (2.5)	37.6 (2.8)	41.3 (2.4)
England	20.8 (2.0)	29.2 (2.3)	53.5 (3.5)	69.6 (2.9)

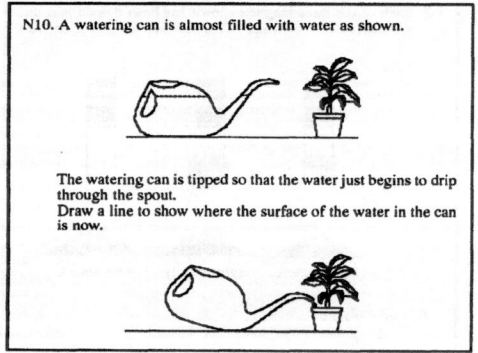

N10. A watering can is almost filled with water as shown.

The watering can is tipped so that the water just begins to drip through the spout.
Draw a line to show where the surface of the water in the can is now.

This particular item is a variation of the classical task used by Piaget to show developmental changes in children. As such, it suggests that maturation can play a role in the pattern of gains. Piaget's comparable task would have argued that there was, in fact, a sudden leap in development between ages nine and thirteen in understanding of the physical world that would cause more children to answer correctly. His argument, however, would have suggested that, if developmental change alone were relevant, virtually all of the older students should have gotten the item correct. Certainly development, especially taking into account gain patterns for other items, offers no explanation for why the percent correct for US students should consistently be below the cross-national means at all four grades. No one has even postulated developmental differences this widespread among children in different countries or as consistent over time.

Exhibit 5.15. *Results for a Life Science Item Appearing on Tests for Third, Fourth, Seventh, and Eighth Grade Students.*

	Grade 3		Grade 4		Grade 7		Grade 8	
USA	29.9	(2.0)	37.3	(1.9)	55.0	(2.2)	58.9	(1.6)
International	*40.1*		*47.7*		*69.8*		*72.6*	
Australia	32.8	(2.9)	37.8	(2.5)	64.5	(1.5)	69.2	(1.1)
Canada	23.8	(2.0)	32.7	(2.7)	57.0	(1.9)	55.9	(1.5)
Czech Republic	65.3	(2.5)	78.8	(2.0)	93.7	(1.0)	95.3	(0.7)
Hong Kong	46.5	(2.4)	62.1	(2.7)	75.1	(1.9)	78.6	(1.6)
Hungary	44.5	(2.6)	65.3	(2.5)	81.3	(1.5)	88.1	(1.1)
Israel			44.8	(3.6)			61.0	(2.2)
Japan	59.9	(2.3)	68.4	(2.1)	79.6	(0.9)	82.3	(1.0)
Korea	50.9	(3.0)	55.2	(2.7)	75.2	(1.5)	81.8	(1.2)
Netherlands	40.0	(2.5)	46.0	(3.6)	71.7	(3.2)	74.1	(2.3)
New Zealand	31.1	(3.2)	33.4	(3.0)	57.2	(1.9)	68.0	(1.8)
Norway	27.8	(3.2)	33.1	(3.2)	61.2	(2.4)	66.6	(1.7)
Singapore	26.1	(1.9)	63.5	(1.7)	87.5	(1.5)	87.3	(0.9)
Thailand	51.5	(3.7)	69.0	(3.3)	82.8	(1.5)	82.9	(1.4)
England	28.6	(2.6)	34.7	(2.7)	53.4	(2.1)	62.1	(2.1)

P9. Seeds develop from which part of the plant?

A. Flower

B. Leaf

C. Root

D. Stem

Exhibit 5.15 presents a third science item, this time from the life sciences, specifically botany. There is a gain of about seven percent from third to fourth grades, of about 18 percent from fourth to seventh, and almost four percent from seventh to eighth. This pattern is roughly like that of the item in Exhibit 5.14. Again, US achievement was consistently below the cross-national means at all grade levels.

A human biology item, shown in Exhibit 5.16 on page 156, is another item that appeared on the tests for both nine- and thirteen-year-olds. It dealt with human nutrition and the value of fruits and certain vegetables in terms of minerals, carbohydrates, etc. About half of the US third grade students could answer this item correctly and there was a difference of over 11 percent between third and fourth grades. However, there was a minimal difference between fourth and seventh grades along with more modest gains (under eight percent) from seventh to eighth grades. Since this item involves a number of specific terms ('carbohydrates', 'protein', 'minerals', 'vitamins', 'water content'), it is unlikely that these performances could be explained by life experience, especially the achievement of the nine-year-olds. The lack of significant growth from fourth to seventh grades suggests that this item was in US science

Exhibit 5.16. *Results for a Human Biology Item Appearing on Tests for Third, Fourth, Seventh, and Eighth Grade Students.*

	Grade 3		Grade 4		Grade 7		Grade 8	
USA	50.0	(3.3)	61.8	(1.9)	62.7	(3.3)	70.4	(1.9)
International	59.1		66.4		74.3		75.6	
Australia	47.3	(3.1)	57.3	(2.7)	64.4	(2.7)	67.6	(1.8)
Canada	49.5	(2.3)	58.4	(3.4)	66.1	(2.8)	67.9	(2.4)
Czech Republic	74.8	(2.1)	82.8	(2.1)	90.4	(1.5)	92.5	(1.5)
Hong Kong	74.5	(2.1)	74.1	(2.6)	72.2	(2.0)	67.2	(2.3)
Hungary	65.5	(2.9)	82.2	(2.3)	93.9	(1.3)	93.1	(1.5)
Israel			72.1	(3.5)			86.9	(2.0)
Japan	52.7	(2.1)	64.0	(1.7)	80.2	(1.4)	87.4	(1.2)
Korea	67.8	(2.9)	78.5	(2.1)	75.7	(2.3)	81.4	(2.8)
Netherlands	91.3	(1.7)	92.5	(1.7)	87.4	(2.5)	84.5	(3.2)
New Zealand	47.7	(3.2)	55.3	(3.2)	63.8	(2.6)	69.9	(2.1)
Norway	58.7	(3.3)	77.0	(2.7)	80.7	(3.0)	81.0	(2.3)
Singapore	62.0	(2.0)	71.6	(1.9)	81.6	(2.2)	86.8	(1.5)
Thailand	40.3	(3.0)	44.8	(3.7)	76.6	(2.0)	80.6	(1.9)
England	59.3	(2.7)	58.5	(2.7)	61.2	(3.7)	65.6	(3.4)

> I10. What is the BEST reason for including fruits and leafy vegetables in a healthy diet?
>
> A.   They have a high water content.
>
> B.   They are the best source of protein.
>
> C.   They are rich in minerals and vitamins.
>
> D.   They are the best source of carbohydrates.

curricula in early grades, but not in the later grades of elementary school. At every grade, US students scored below the international average although, more so at third and seventh grades.

Exhibit 5.17 changes the focus to an earth science item. While this item does not involve technical terms, it is a good example of an item that involves 'academic' school science since knowledge that would allow students to answer this item correctly is unlikely to have come from life experience or even a hands-on investigation. US students consistently scored above the international average by six to 14 percent, the larger differences coming in the earlier grades. A surprising 83 percent of US third graders answered this item correctly. There were gains across the four grades (although very little between seventh and eighth grades), but the gains were less for the two adjacent grade levels than they were from fourth to seventh grade. The academic nature of the task, the early high achievement, the consistent growth, and the narrowing of the gap between the US and the international average all suggest that such 'knowledge receiving' or 'academic' tasks were strong points for US students and more suited to the nature of the school science they experienced than was true for earlier grades in other countries.

Exhibit 5.17. *Results for an Earth Science Item Appearing on Tests for Third, Fourth, Seventh, and Eighth Grade Students.*

	Grade 3		Grade 4		Grade 7		Grade 8	
USA	83.0	(1.3)	86.3	(1.1)	91.0	(0.8)	92.1	(0.6)
International	69.4		76.2		85.1		86.4	
Australia	70.9	(1.7)	75.3	(1.2)	84.0	(1.3)	86.9	(0.8)
Canada	77.1	(1.2)	83.9	(1.1)	90.2	(0.9)	91.0	(0.9)
Czech Republic	69.6	(1.5)	75.5	(1.5)	91.1	(1.0)	96.9	(0.5)
England	77.8	(1.4)	83.9	(1.3)	89.8	(1.2)	90.7	(0.9)
Hong Kong	69.8	(1.5)	82.0	(1.3)	87.3	(1.2)	86.8	(1.2)
Hungary	61.8	(1.9)	78.4	(1.4)	91.8	(0.7)	93.4	(0.7)
Israel			65.3	(1.7)			75.3	(2.4)
Japan	83.5	(1.0)	84.2	(0.9)	86.6	(0.7)	86.6	(0.7)
Korea	81.0	(1.2)	83.9	(1.3)	78.5	(1.3)	84.1	(1.1)
Netherlands	72.3	(1.8)	82.3	(1.4)	89.3	(0.9)	91.7	(1.2)
New Zealand	72.2	(1.7)	79.8	(1.7)	89.1	(1.1)	89.9	(0.8)
Norway	68.5	(1.8)	82.7	(1.6)	91.6	(1.2)	92.8	(0.7)
Singapore	72.0	(1.0)	72.4	(1.0)	78.5	(1.3)	83.6	(1.3)
Thailand	59.3	(2.3)	61.4	(1.8)	65.7	(1.5)	79.9	(1.4)

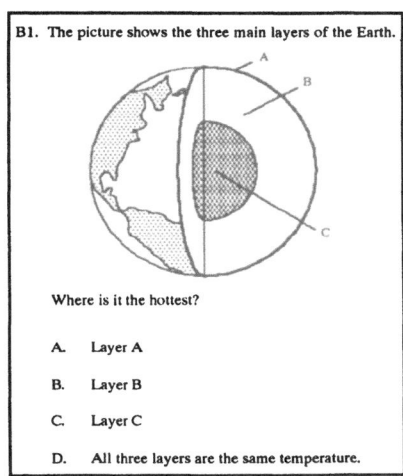

B1. The picture shows the three main layers of the Earth.

Where is it the hottest?

A.   Layer A

B.   Layer B

C.   Layer C

D.   All three layers are the same temperature.

Returning to mathematics items, Exhibit 5.18 presents a common fraction task that, because of its 'short answer, open response' format demands some understanding of how to compare common fractions and of the relative roles of either numerators, denominators, or both in determining the comparative sizes of fractions. Again, US students performed better than the international average at every grade level but less so at the two higher grades. Even by third grade, over half of the US students correctly responded to this item and there were consistent differences across the grades, especially from third to fourth grades and only slightly less so for the three grade span from fourth to seventh.

Perhaps this indicates the importance of curricular coverage fairly directly since common fractions were an early topic in most US mathematics curricula and continued to be so. This was somewhat less true in many other TIMSS countries.

Exhibit 5.18. *Results for a Common Fractions Item Appearing on Tests for Third, Fourth, Seventh, and Eighth Grade Students.*

	Grade 3		Grade 4		Grade 7		Grade 8	
USA	56.1	(2.7)	68.3	(1.4)	79.0	(2.2)	81.0	(1.9)
International	*41.2*		*57.4*		*74.0*		*75.8*	
Australia	43.7	(2.9)	58.8	(1.9)	76.2	(2.3)	77.9	(1.6)
Canada	45.8	(2.4)	61.9	(3.1)	74.5	(2.4)	79.9	(1.6)
Czech Republic	30.5	(2.2)	59.5	(1.9)	81.1	(2.2)	83.2	(2.1)
England	45.0	(2.0)	55.0	(2.0)	78.6	(3.1)	79.3	(2.6)
Hong Kong	74.3	(1.9)	83.9	(1.3)	86.0	(2.2)	84.9	(2.2)
Hungary	38.1	(2.4)	63.2	(1.9)	84.9	(2.0)	86.9	(1.9)
Israel			65.0	(2.6)			79.9	(3.1)
Japan	80.6	(1.5)	78.3	(1.5)	85.3	(1.3)	87.0	(1.2)
Korea	72.3	(2.2)	90.1	(1.3)	76.7	(2.3)	83.9	(2.2)
Netherlands	31.6	(1.7)	49.7	(2.3)	86.1	(2.5)	76.2	(3.3)
New Zealand	38.5	(2.3)	51.3	(2.6)	81.2	(2.4)	80.1	(2.0)
Norway	16.6	(2.0)	38.3	(2.6)	72.8	(5.3)	84.4	(1.6)
Singapore	62.2	(1.3)	76.2	(1.2)	84.4	(2.1)	87.9	(1.6)
Thailand	55.4	(3.8)	72.9	(2.3)	67.6	(2.3)	73.4	(2.1)

16. Write a fraction that is larger than $\frac{2}{7}$.

Answer: _____

Exhibit 5.19 is likely another indication of curricular coverage, as well as mathematics curricula that become undemanding by eighth grade. The item deals with a whole number operation, but does involve the complication of 'repeated borrowing' because of the three zeros in 6000. US performance at each of the four grade levels shadows the international average fairly closely, typically a few points above it in every grade except fourth. There were large differences (both in the US and internationally) between third and fourth grades and between fourth and seventh grades. There were minimal differences between seventh and eighth grades. This seems clearly to reflect an effect of curricular attention, both in the US and internationally. In the US and some other TIMSS countries, content related to this item remained in the curriculum for many eighth graders. The small differences between seventh and eighth grades indicate the low marginal utility of such continued exposure to this content and, especially in some of the other TIMSS countries, may reflect only experience rather than curricular coverage.

The general purport of all specific items discussed above is that, while developmental and life experience factors may be involved in accounting for

Exhibit 5.19. *Results for a Whole Number Item Appearing on Tests for Third, Fourth, Seventh, and Eighth Grade Students.*

	Grade 3		Grade 4		Grade 7		Grade 8	
USA	52.0	(3.0)	70.7	(2.2)	88.1	(2.1)	90.5	(1.1)
International	50.1		71.1		85.7		86.1	
Australia	22.7	(2.6)	46.6	(2.3)	82.2	(2.4)	81.5	(1.7)
Canada	37.9	(2.9)	61.2	(3.1)	90.7	(1.6)	91.2	(1.7)
Czech Republic	60.8	(2.6)	83.3	(1.8)	96.8	(1.1)	97.4	(0.9)
England	23.0	(2.4)	36.0	(2.5)	58.6	(3.2)	65.3	(3.2)
Hong Kong	78.4	(2.3)	88.6	(1.5)	89.6	(1.4)	88.9	(1.9)
Hungary	73.5	(2.5)	90.8	(1.5)	95.3	(1.3)	95.8	(1.2)
Israel			71.4	(3.4)			95.3	(1.4)
Japan	72.6	(1.9)	89.0	(1.4)	89.3	(1.4)	92.5	(1.2)
Korea	87.6	(2.1)	92.8	(1.6)	90.6	(1.6)	89.4	(1.8)
Netherlands	45.5	(3.1)	86.4	(2.0)	87.6	(2.6)	81.8	(3.6)
New Zealand	15.2	(2.3)	29.5	(3.6)	69.4	(3.5)	71.0	(2.3)
Norway	9.5	(1.9)	59.6	(3.7)	85.1	(5.5)	86.9	(2.0)
Singapore	90.1	(1.0)	91.5	(1.0)	98.3	(0.6)	97.7	(0.7)
Thailand	49.8	(3.8)	64.9	(3.0)	86.5	(1.6)	86.2	(1.6)

R12. Subtract: 6000
                   − 2369

A. 4369
B. 3742
C. 3631
D. 3531

achievement changes, curricular factors undoubtedly are. More direct studies of these linkages are underway and will be presented elsewhere. For now, the main evidential value of examining these link items is that their differences rule out explanations based on factors such as maturation, life experience, or some general measure of mathematics or science achievement. These items require specific knowledge and understanding. General levels of attainment in mathematics or science may reflect part of the explanation for differences in performance, but they are not sufficient to explain the differences in performance among items. If these general measures are claimed as the sole indicators of achievement and, because of their apparent lack of correlation with curriculum measures, used to dismiss the relevance of curricular factors, this is an abuse of their nature as aggregates and scaled scores that have kept items that fit the model and dismissed as 'error' anything too complex. If general achievement measures are used in combination with more specific measures together with estimates of achievement gains, more accurate and sensitive explanations of student performances result. When this happens, curriculum effects enter almost every explanation regardless of content area, item, or country.

Assessing the Achievement Gains
of US Students: A Summary

The ability to estimate achievement gains for a single grade allowed us to extend our consideration of whether curriculum is related to achievement in ways that provide even stronger arguments for the proposition that 'curriculum matters' than was the case when considering only the comparative status of US students on more specific content areas and items. The following summarize and re-state some of the points we sought to make.

- Since two adjacent grades were tested for both nine- and thirteen-year-olds, country-level gains between the grades can be estimated. When we measure gains in terms of increased percentages of those who answered an item correctly, the gains for the US nine- and thirteen-year-olds in both mathematics and science were similarly small very for most of the different topic areas.

- The comparable gains of other countries for nine- and thirteen-year-olds in both mathematics and science were more varied with noticeable peaks for some topics, compared to the US' relatively flat gains profile.

- In general, countries varied relative to other nations more in terms of their gains than their comparative status. This makes it even clearer that differences in gains among topics were especially likely to have reflected curricular differences and that 'curriculum matters.' *Status* was determined by performance based on aggregate learning over many grades, while *gains* were based on learning within a single grade. Were it to be status that had showed the greater variability among topics, it might suggest that differences were artifacts of the measurement process and less likely to reflect other than the most marked curriculum differences. Since the greater variability was for the more curricularly-sensitive measure of gains, this suggests that intra-topic variance was related to curriculum rather (as well as, perhaps) than to other factors such as maturation and life experience.

- The case of the US, while disappointing for its citizens, becomes interesting in the context of this reasoning since it illustrates relatively low gains that were fairly consistent across topics. These are outcomes one would expect from a highly redundant curriculum with evenly distributed but shallow coverage of most topics in any one grade – which is what is meant by calling US mathematics and science curricula 'a mile wide and an inch deep.'

- Other countries' achievement variance allowed most countries to be both among the best for some topics and among the worst for others. The comforting argument that it is better to achieve consistent gains across many topics, rather than marked highs and lows, is disallowed by the low cumulative status among other TIMSS countries which accompanies this US consistency. Consistently high scores might be comforting. Consistently mediocre scores are not.

- The nine- and thirteen-year-olds' test results should give us no comfort since they make it clear that the 'you can't have it all' adage is borne out by TIMSS countries' gains across topics. Virtually no country showed top gains in all topic areas. Most showed large gains in some areas, often reflecting topic emphases in that country for that particular grade. Comparing our fourth grade results to our eighth grade results shows that aggregating small gains, though consistent across topics, are related to comparatively disappointing status among the TIMSS countries. In particular, our comparatively poorer showing at the eighth grade than at the fourth grade suggests that consistent but small gains did not gradually surpass the cumulative gains of countries with highs and lows – especially if those highs and lows reflected curricula carefully orchestrated to eventually emphasize all central topics.

- From the results of our fourth grade and eighth grade gains, it is clear that through eighth grade we have no 'highs and lows', no profile of strengths and weaknesses as other countries do (even those whose status is among the poorest).

Concluding Remarks: Curriculum Does Matter

This chapter's analyses of achievement seem both revealing and important in avoiding the misuse of cross-national achievement comparisons. First, they demonstrate fundamentally that 'a test is not a test' or, perhaps better, that 'all tests are not created equal.' Comparative performances on achievement tests in mathematics and science, including the TIMSS tests, depend on how specific the contents are. Highly aggregated and scaled scores tend to focus on the commonalities between disparate items, to suggest that 'mathematics' or 'science' is monolithic, and to produce achievement measures that are relatively insensitive to curricular differences. If this artifact of aggregation and scaling is used to dismiss the importance of curriculum, it is a serious case of ignoring a method's limits.

As we considered specific areas, we discovered that no country could 'do it all and do it all well.' Most countries had content areas with stronger and weaker comparative achievements, even countries that fared poorly overall. These differences seemed linked to curricular and educational systemic differences and it is important to understand this in order to have a better basis for deliberate change and improvement.

These analyses, especially those of achievement gains, particularly called into question the practice common to most US education systems of using curricula that deliver consistently small amounts of instructional attention to large numbers of topics. This was associated with estimated gains that were consistent but did not accumulate by eighth grade into satisfactory comparative attainments in either mathematics or science. This certainly poses questions about our curricular practices and their relation to cumulative attainment that will be more fully answered when the TIMSS end-of-secondary results are more fully analyzed (see the last chapter below).

The data presented and analyzed here seem to reflect patterns that likely are related to typical American educational practices. Systemic factors not only shape the nature of mathematics and science education, but they almost certainly have consequences for the attainments of our children in these subjects, assuming schooling and curriculum really do matter to what children achieve.

Notes–

[1] Wolfe, R. G. (1989, February). *An indifference to differences: Problems with the IAEP-88 study*. Paper presented at the NSF-sponsored Research Conference on the SIMS Data, University of Illinois, Champaign, IL.

[2] International means are the unweighted means based on all countries participating in the TIMSS assessment at each of the indicated student grade levels, i.e., 26 countries at grade four and 41 countries at grade eight. The significant difference categories for each scale were determined at the .05 significance level adjusted for multiple comparisons. Standard errors for these comparisons were determined using the jackknife procedure. Only the focal countries for this report are included in the exhibits. Therefore, each significant difference category may contain additional countries to those included in the exhibit tables.

[3] Schmidt, W. H., Jakwerth, P. M., & McKnight, C. C. (in press). Curriculum-sensitive assessment: Content does make a difference. *International Journal of Educational Research*.

[4] For illustrative purposes, all countries have been included in these displays and a few other comparisons in this chapter and chapter six. For most displays we chose to focus on the more limited set of countries identified in chapter one.

[5] See: Schmidt, W. H., McKnight, C., Valverde, G. A., Houang, R. T., & Wiley, D. E. (1997). *Many Visions, Many Aims, Volume I: A Cross-National Investigation of Curricular Intentions in School Mathematics*. Dordrecht/Boston/London: Kluwer; and Schmidt, W. H., Raizen, S. A., Britton, E. D., Bianchi, L. J., & Wolfe, R. G. (1997). *Many Visions, Many Aims, Volume II: A Cross-National Investigation of Curricular Intentions in School Science*. Dordrecht/Boston/London: Kluwer.

[6] Schmidt, W. H. (1998, April). *Educational opportunity and achievement in the United States: Summative report from the Third International Mathematics and Science Study*. Paper presented at the American Educational Research Association, San Diego.

[7] Wiley, D. E., & Wolfe, R. G. (1992). Major survey design issues for the IEA Third International Mathematics and Science Study. *Prospects, 22*, 297-304.

Chapter 6
Access to Curriculum Matters

If curriculum does matter, as we argued in Chapter 5, then whether children gain access to the educational possibilities of particular curricula also matters. Some have argued that there is a naturally high variability in the achievement of US children. Instead, we would argue that US children, while they may vary somewhat in how well they make use of the educational possibilities available to them, achieve differently because they are given access to different curricula, curricula that shape and limit what they may achieve in spite of their best efforts. We might label this idea with the phrase 'differential access.' It appears to be a characteristic feature of US education systems, at least in mathematics and science.

This chapter examines the variability of US children's science and mathematics achievements and the relation of that variability to differential access to educational possibilities. Some of these differing possibilities are likely inadvertently related to our characteristic practice of dispersing educational responsibility and decision making to many local and state education systems so that mathematics and science curricula vary among states and even communities. However, part is also intentional either through policies of tracking in mathematics or through specialized topic courses in science. We begin with an analysis of variability in US mathematics and science achievement.

VARIABILITY IN US MATHEMATICS
AND SCIENCE ACHIEVEMENT

Some have argued that one reason why the US has traditionally not done well in cross-national comparisons is because it has such a heterogeneous population. That argument implies that US society, which has been traditionally considered a 'melting pot,' has absorbed so many diverse regional, ethnic, and immigrant elements that there is a resulting social heterogeneity that reveals itself in the great variability of achievement that results from schooling. They would argue that our student population reflects our social heterogeneity and that, somehow, this affects how well students make use of educational possibilities and also shapes the levels of achievement that follow from students' encounters with those possibilities during schooling.

Overall variability. The TIMSS data do not support this argument and suggest quite the opposite. Exhibit 6.1 displays the standard deviation of the scale scores in which global 'mathematics' achievement was reported in previous TIMSS reports. The units of these numbers are artifacts of the scaling process and have no meaning in themselves. We use them here only to compare standard deviations among countries as a measure of variability. The scores for fourth and eighth grade mathematics are each arranged from largest to smallest.

The United States fell towards the middle of both lists. In both cases several countries had higher standard deviations than the US. The counterpart data for science showed only three countries with higher standard deviations than the US at both fourth and eighth grades. This indication of variability in science, especially fourth grade science, occurred for scales that indicated comparatively good US achievement. Our variability was not extreme in any of these cases, but the differences for science bear further consideration.

Essentially, the US was average in its variability at both grades for mathematics and somewhat high in science variability for both grades, relative to other countries. In neither case was our achievement variability particularly extreme. Apparently our perceived social heterogeneity did not translate into unusual achievement heterogeneity when we compared that latter heterogeneity (variability) to its counterparts in other countries.

Surprisingly, there were consistent, moderately strong, positive correlations between mean achievement and variability (here measured as the size of the interquartile range, the spread for the middle 50 percent of students, in units of the scale scores for mathematics and science). Correlations ranged from about .5 to about .7 for both fourth and eighth grades and for both science and mathematics. Countries with higher mean achievement tended to have greater variability among students. This is another way of arguing that, with the exception of fourth grade science, US achievement variability was not unusual. It also required that we examine variability in a way that took into account its relationship to mean achievement.

Variability for higher and lower achieving students. To take a closer look at variability, we calculated the median, the 'width' (score difference) of the second quartile (the twenty-fifth to fiftieth percentile), and the 'width' (score difference) of the third quartile (the fiftieth to seventy-fifth percentile) of student scores. We plotted each of these quartiles against the median for each country in science and mathematics separately and at each population separately. This essentially allowed us to look at separate measures of variability for the lower and upper half of the middle 50 percent of the distribution of each countries' student scale scores. In seeking a systematic relationship between median and these measures of variability, we fitted a regression line

Exhibit 6.1. *Variability for Countries on the TIMSS International Mathematics Scale at Fourth and Eighth Grades*.*

Country	Grade 4	Standard Deviation
Singapore		104
Australia		92
England		91
Greece		90
New Zealand		90
Scotland		89
Hungary		88
Cyprus		86
Czech Republic		86
Ireland		85
United States ● ● ● ● ● ● ●		85
Latvia (LSS)		85
Israel		85
Canada		84
Slovenia		82
Japan		81
Portugal		80
Hong Kong		79
Austria		79
Norway		74
Korea		74
Iceland		72
Netherlands		71
Thailand		70
Iran, Islamic Rep.		69
Kuwait		67

*See note 3 on page 12.

Country	Grade 8	Standard Deviation
Bulgaria		110
Korea		109
Japan		102
Hong Kong		101
Australia		98
Czech Republic		94
England		93
Hungary		93
Ireland		93
Austria		92
Belgium (Fl)		92
Israel		92
Russian Federation		92
Slovak Republic		92
United States ● ● ● ● ● ● ●		91
Germany		90
New Zealand		90
Netherlands		89
Romania		89
Cyprus		88
Greece		88
Singapore		88
Slovenia		88
Switzerland		88
Scotland		87
Belgium (Fr)		86
Canada		86
Thailand		86
Sweden		85
Denmark		84
Norway		84
Latvia (LSS)		82
Lithuania		80
France		76
Iceland		76
Spain		73
South Africa		65
Colombia		64
Portugal		64
Iran, Islamic Rep.		59
Kuwait		58

to each plot of data.[1] Exhibit 6.2 presents these data for fourth and eighth grade mathematics scores. Exhibit 6.3 presents them for fourth and eighth grade science scores.

In eighth grade mathematics, there was a strong positive relationship between countries' median achievement scores and the variability (spread) in the lower half of the middle 50 percent of that countries distribution. The US variability was consistent with its predicted value, differing only slightly as Exhibit 6.2 shows (comparing the circled point with the corresponding point on the regression line). For fourth grade (or its equivalent) mathematics, the relationship was not so strong (as can be seen in the smaller slope of the regression line). However, the US variability was again consistent with its predicted value.

Exhibit 6.2. *Median Scores versus Second and Third Quartile Scores for*
Mathematics Scale Scores for Fourth and Eighth Grades.

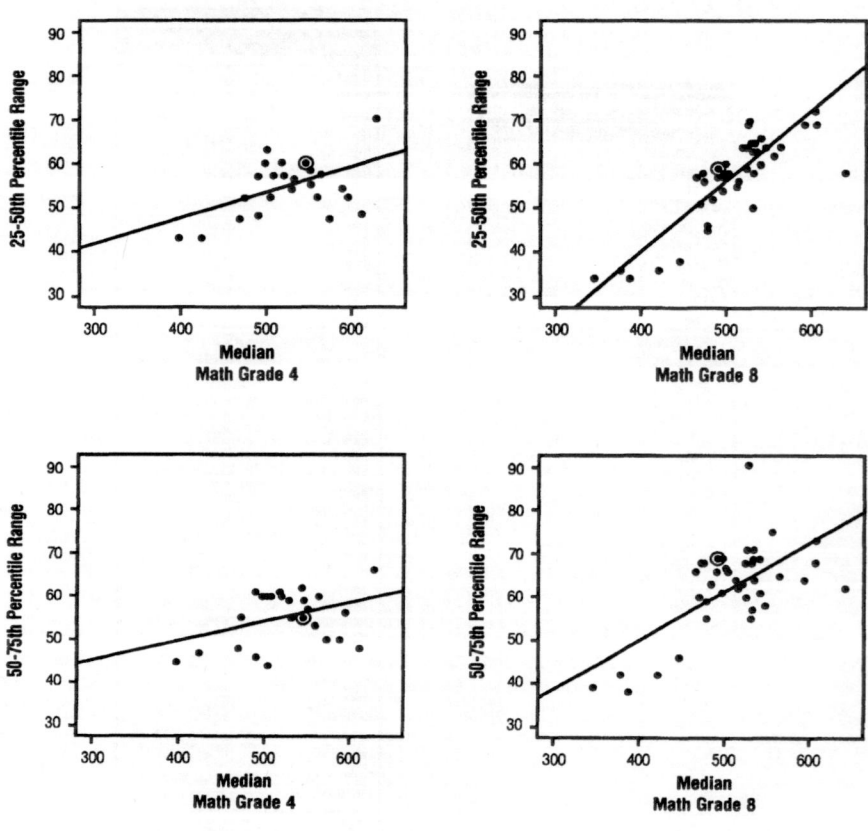

This suggests that US 'lower middle' (twenty-fifth to fiftieth percentile) variability was consistent with that of other countries, taking into account the relationship between variability and median achievement.

The case for 'upper middle' (fiftieth to seventy-fifth percentile) variability
was somewhat different. There was again a relationship between variability
and median achievement, more so at eighth grade than at fourth (as can be seen
from the slopes of the regression lines). At fourth grade, our variability was
consistent with that of other countries with a similar median level of performance. This was not true at eighth grade for which US variability was considerably higher than was predicted from its achievement median.

Why was variability higher than expected for the 'upper middle' group of
students in US eighth grade mathematics but not for the 'lower middle' group,
or for either group, in fourth grade mathematics? One hypothesis is that this

Exhibit 6.3. Median Scores versus Second and Third Quartile Scores for Science Scale Scores for Fourth and Eighth Grades.

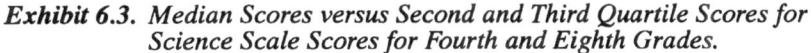

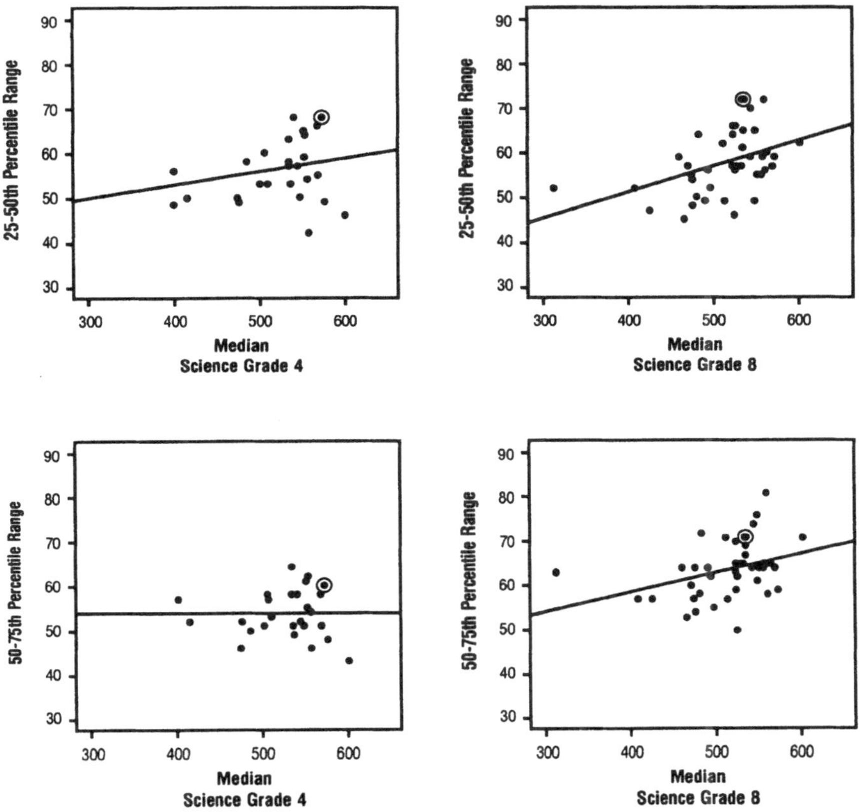

greater variability is a curricular effect from tracking which affects students that perform above the median for US eighth grade mathematics. We discussed earlier in this report how the content covered and emphasized varies by tracks. Most US students with below-median achievement were likely to have been in general mathematics classes (or remedial classes that did not differ significantly from them). US eighth grade students in enriched and algebra classes were likely to have come almost exclusively from those that had above-median achievement. Thus, the mathematics curricula to which students with achievement above the median were exposed likely differed more (coming from different types) than was true for students with below-median achievement. If access to educational possibilities through curricula affect achievement, differing access to curricula should result in greater achievement variations.

In eighth grade science, the relationships between median achievement and both 'upper middle' and 'lower middle' variability were less strong than in mathematics. At fourth grade, there was essentially little or no relationship between median achievement and either kind of variability(the slopes of the regression lines being near zero).

In US eighth grade science, both the 'upper middle' and 'lower middle' variability was greater than would be predicted from US median achievement. Earlier discussions of curricular differences in eighth grade science point out that the main differences were through discipline-based courses (earth science, physical science, life science) and a more eclectic general science. The result of such course-type differences is that there were distinctly different science curricula to which US eighth grade students could have been given access. Since these courses differed by orientation rather than from trying to produce more locally homogeneous groups of students in tracks (as was the case for mathematics), they were likely to affect both the lower and upper parts of the achievement distribution on the TIMSS test which assessed all science areas, and were not likely to affect just the upper part (as would be the case for tracking in mathematics). The findings on greater than expected variability in US eighth grade science were thus consistent with an explanation in terms of differential access to curricula with different profiles of content covered and emphasized.

There was also greater variability than expected based on median achievement in the 'lower middle' and 'upper middle' groups for fourth grade science. An explanation similar to that given for eighth grade is not reasonable here since similar discipline-based curricular differences did not exist at the fourth grade level. However, other curricular differences might help in understanding this greater variability. In Chapter 2 we discussed the variation in the amount of time spent on science across US classrooms. This varied from an average of about 10 to 60 minutes per day. Such differences in time spent could be related to greater achievement variability, especially when coupled with the fact that fewer topics were commonly taught by 90 percent or more of US fourth grade science teachers. Only two topics met this 90 percent taught criterion for fourth grade science compared to seven for fourth grade mathematics teachers. The result might be greater variability in what topics were taught to which students.

Another possibility is that many of the items could be answered from life experience (as discussed at the end of Chapter 5). They might also be answered from out-of-school science learning through television and other media. Either of these possibilities could have introduced further variation among students in their science achievement since such experiences would not be uniform. These explanations are at least plausible, although other explanations are also viable. The actual explanation, whether a combination of these or some other factor altogether, must remain conjectural.

Are our best among the world's best? An argument is often made that, regardless of mean levels of achievement, our best students achieve in ways

Exhibit 6.4. *Boxplots Portraying the Distribution of Students' Mathematics Score at Fourth and Eighth Grade.*

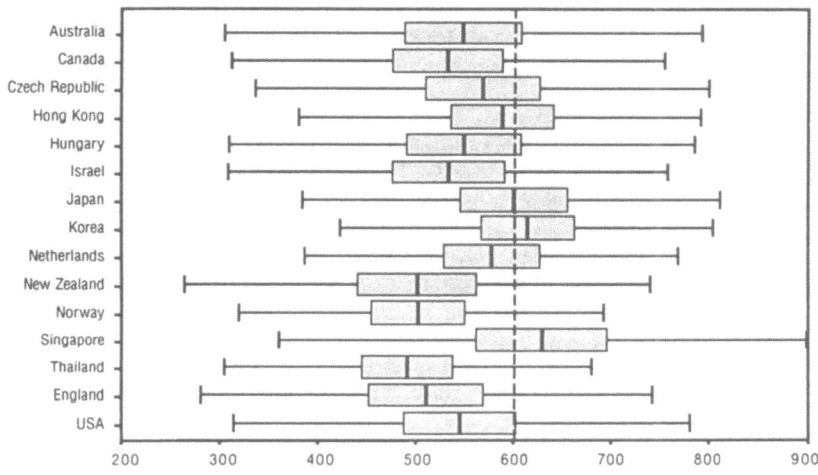

Fourth Grade Mathematics Score

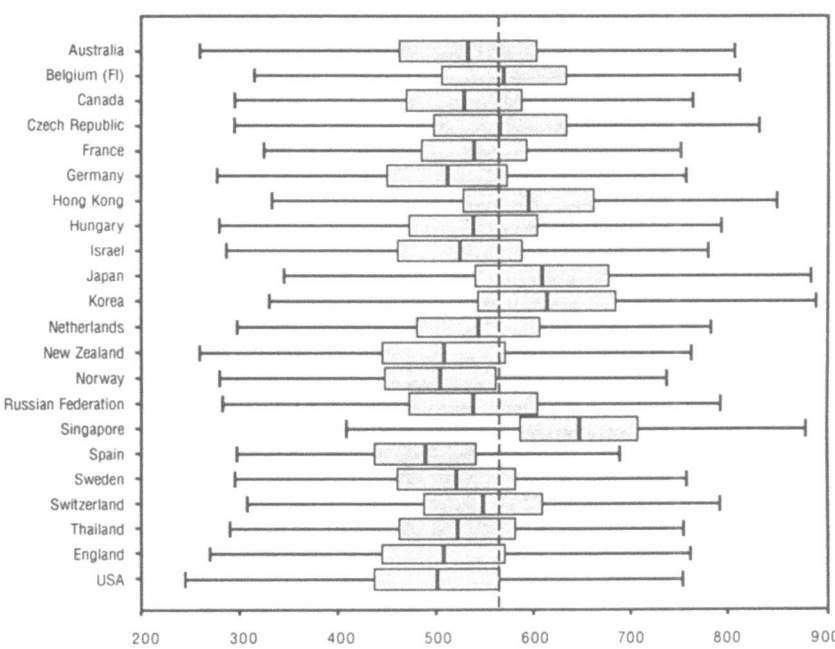

Eighth Grade Mathematics Score

Exhibit 6.5. *Boxplots Portraying the Distribution of Students' Science Score at Fourth and Eighth Grade.*

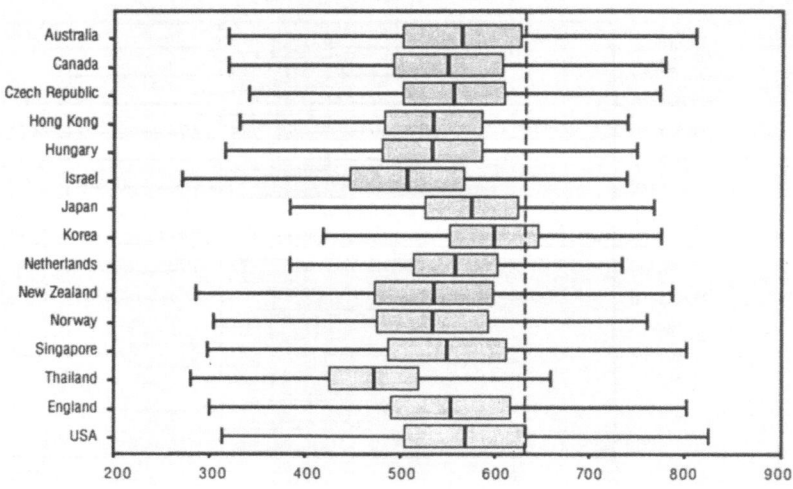

Fourth Grade Science Score

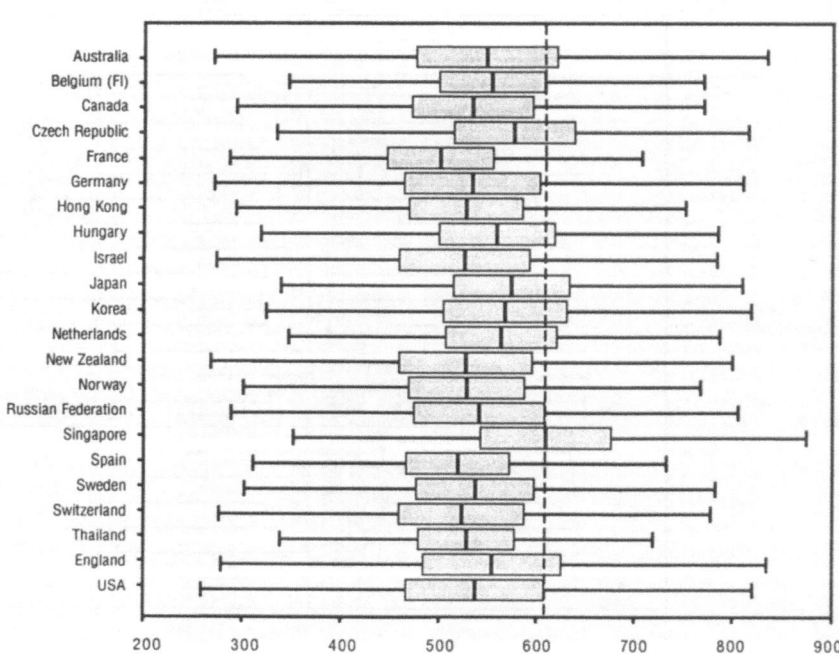

Eighth Grade Science Score

comparable to the best students in other countries. Variability data which display the entire distribution of achievement scores help to determine whether or not this argument is valid.

Exhibit 6.4 shows the distribution of overall scale score achievement for fourth and eighth grade mathematics. Exhibit 6.5 shows similar distribution data for fourth and eighth grade science. From Exhibit 6.4 it is clear that our seventy-fifth percentile (high end of the box) fell below the seventy-fifth percentile of 19 of the 21 countries focused on in this report in eighth grade mathematics. Our best fell below the twenty-fifth percentile for Singapore and just above the twenty-fifth percentile in Japan, Korea, and Hong Kong. Our best would only be slightly above average in the Czech Republic, France, the Netherlands, and the Russian Federation. For fourth grade mathematics, the display makes it clear that the story was only slightly better.

In similar comparisons regarding science, we fared much better, especially at the fourth grade level. At eighth grade, our seventy-fifth percentile was below the seventy-fifth percentile of nine of the 21 countries. The US seventy-fifth percentile was essentially the median in Singapore. We also had greater variability in our science achievement which made it more likely that the upper end of our distribution fell comparably higher in relation to the distributions of other countries. This greater variability was likely induced by curricular differences and other factors already discussed. In any case, there is little support for the claim that our best are among the best in the world (or at least in TIMSS countries), especially for mathematics.

SOURCES OF VARIABILITY IN US MATHEMATICS AND SCIENCE ACHIEVEMENT

Achievement variability may stem from curricular differences as we have already discussed. It may also come from regional and other differences created by our characteristic pattern of having independent local and state education systems. It is thus important that we take a closer look at the sources of variability in US science and mathematics achievement.

American practices in the organizing of curricula and schooling in science and mathematics have among their consequences differential access to educational possibilities and are thus likely related to achievement differences. Certainly our system of distributed decision making among state and local education systems produces inadvertent differences in access to activities aimed at learning specific mathematics and science. Our widespread policies of tracking in mathematics and of discipline-oriented courses in science, even done with the best of intentions, are also likely to produce differing access to educational possibilities. We might say that the US approach to education is structured in ways that tend to increase or create differences in access to specific science and mathematics curricular contents. This differential access seems likely to be linked to differences in achievement and thus to achievement variability.

This argument about differential access being related to greater variability is not inconsistent with our earlier argument that total US variability was not out of line with that of other countries, especially when achievement levels (systematically related to variability) are taken into account. Here we are discussing the structure and sources of achievement variability in the US. We are claiming that basic structural and systemic features produce differential access. This likely would be reflected in variation due to differences among schools, among tracks within schools, and among classes within tracks if our hypotheses about differential access is correct. To explore this we need to partition the variation in US achievement scores into components of school differences, class differences, and differences among students. The statistical technique for doing this is variance decomposition and it yields estimates of the proportions of variance due to different sources.

Most other countries, by explicit policies, commonly structure themselves by means of a common curriculum through eighth or ninth grades to attain broad commonalities in achievement among schools and classes within schools. As a result, most of the variance in achievement scores for these countries is due to differences among students(which are created by differences in ability, motivation, engagement, and many other factors). This is quite different from what we hypothesize to be true for the US. It is also different for several countries (e.g., Germany and the Netherlands) which, although they had a common curriculum, tracked students by general ability into different schools.

Before we present the results of an analysis of variance components, certain caveats must be stated. First, to separate differences among schools from differences among classrooms within schools, sampling needs to be designed to include two classrooms at the same grade level within each school. The US sample did this and we can analyze school differences and classroom differences as separate sources of overall variability. Most TIMSS countries (other than Australia) did not do this. As a result, differences among schools and differences among classrooms are confounded for those countries. We can still distinguish variance due to the conflated school and classroom differences from variance due to differences among students, but we cannot separate variance due to schools from variance due to classrooms.

Second, there is a complication if we extend the US variance component analysis to cover tracking. Tracking, given the important role it plays in US mathematics curricula, should have been a stratification factor in designing sampling for US eighth grade mathematics classrooms, but it was not. As a result, differences among classrooms within schools confound differences among tracks within schools with differences among classrooms within tracks. We have tried to adjust for this by doing separate analyses given the specific tracking structure within each school. However, this is not the ideal way to examine variation due to tracking.

In eighth grade mathematics, this results in both an overall analysis and one for schools with different types of classrooms in the sample. For example, we did a variance decomposition for all schools which only had general mathematics classes sampled. The classroom component of variance in such a case reflects differences among classes within the general mathematics track. We did similar analyses of schools for which both sampled classes were pre-algebra or both algebra. These give estimates of differences among classrooms within these tracks. However, these estimates are based on different sets of schools. A better sampling design would have permitted an analysis of classrooms within tracks for all tracks while using the same set of schools.

Other analyses were done when one classroom within a sampled school was in the algebra track and the other in a different track, for instance general mathematics. Here, the classroom component of variance is an estimate that confounds track differences with classroom differences within tracks. This was done for various combinations of three tracks: general mathematics, pre-algebra and enriched mathematics, and algebra classes. With these caveats in mind, let us turn to the results of the variance component (source) analysis.

Comparative variance components for global scale scores. Many countries did attain their policy goal of equal access to learning. Using the total score in mathematics at eighth grade or its equivalent, the component of achievement variation reflecting the combined effect of school and class-within-school differences was very small, accounting for less than 30 percent of the total achievement variance in 10 out of the 22 countries. The remaining 70 percent or more of achievement variance would be attributed to differences among students. Thus, the vast majority of variance came from student factors and only a comparatively small proportion from structural differences that affected individual student achievements in ways over which they had no control. In Japan, Korea, Norway, and Sweden the proportion of the total score variance attributable to schools and classrooms was around 10 percent.

In the US, only about half of the achievement variance in mathematics could be attributed to differences among students within classes. Only in five countries, Belgium (Flemish), Hong Kong, Switzerland, Australia, and Germany, was the student level variance around the same size as that in the US. In Germany, much of the school/classroom level variance may be attributed to the heavy tracking students experience as they attend different types of schools. Only in the Netherlands, where students are also heavily tracked by school, is the percentage of variance for students appreciably less than that for the US.

These data are consistent with our earlier hypotheses. Structural features in US education systems exaggerate differences in access among schools and among classrooms within schools (because of tracking) resulting in a large percentage of achievement variability related to these structural features.

Exhibit 6.6. Sources of Variance for the Eighth Grade International
Mathematics Score.

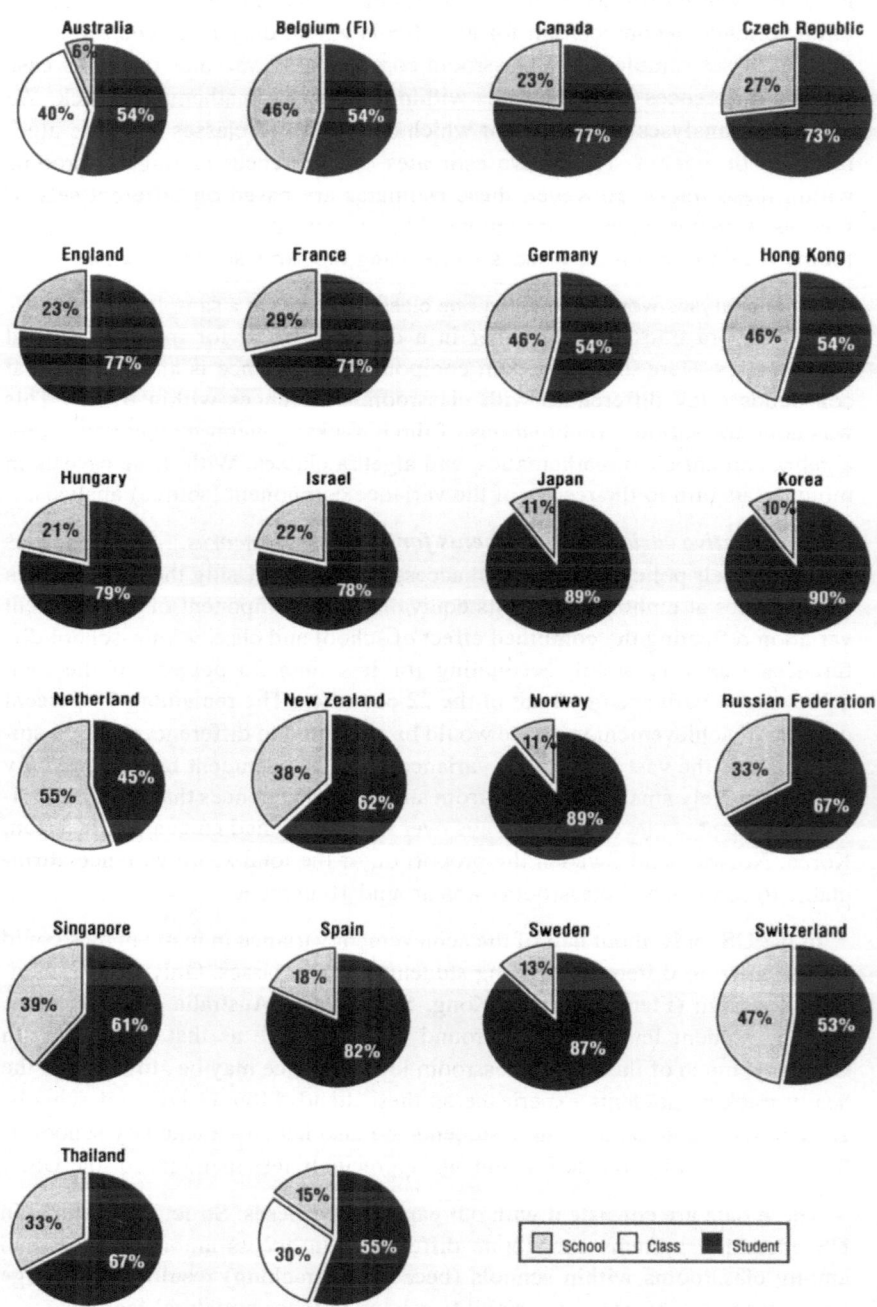

These findings are about differing access to learning possibilities across the country and its relationship to student achievement. We cannot tell from these analyses whether the differences in achievement across schools and classrooms within schools follow from differences in access to educational possibilities, differences in the process by which students come to be in a particular track in mathematics, or some combination of the two. What is clear is that structural features of US educational systems and of how education is organized in the US had profound implications for variations in how much mathematics and science was learned.

Detailed analyses of US variance components. The analyses so far were done to compare the US in a fashion that could be consistent across other TIMSS countries. In the next set of analyses, we use detailed sampling considerations from the US sampling plan and imputed data at the item level which we can do for the US but not easily for other countries. This allows us to characterize the variance components for the US more completely.

We used the same US achievement data in the form of correct and incorrect answers to items, but we did not use a scaled version of them to form a total test score.[2] Here, we work only with achievement in eighth grade mathematics because we want to examine the effects of tracking more precisely. The results are consistent with the previous analyses using scaled scores, but the scaling appears to lead to possible underestimation of the variance components attributable to schools and classrooms.

Using unscaled, imputed total scores leads to a pool of variance of which only about 25 percent is attributable to differences among students in classrooms (compared to about 50 percent for scaled scores in eighth grade mathematics).[3] This means that 75 percent of the variance in these eighth grade mathematics total scores are attributable to school (about 25 percent) and classroom-within-school (about 50 percent) differences. This estimate implies that the structural features previously discussed had an even more pronounced impact on mathematics achievement variability than suggested by our earlier estimates. Minimally, the two estimates provide an upper and lower bound on the percentage of variation attributable to individual differences.

These data also allow a richer and more detailed examination of tracking issues. First, it is possible to explore the effects of differences among tracks independently of the effects of classrooms within tracks. As we said above, for the unscaled total score, the total effect of differences among classrooms within tracks is about 25 percent. However, with a new possibility for examining variance due to differences among tracks, the variance attributable to differences among schools shrinks to about 25 percent of the achievement variance for the total unscaled eighth grade mathematics score. The remainder is attributable to differences among tracks. If we base our estimates only on the sub-sample of schools in which the sampled classes were from different tracks, we get an estimate of combined track and classroom effects which is almost 80 percent of the total score's variance. Regardless of the particular combinations of tracks we chose for analyses, the sources of variance were about equal across school, track, classroom-within-track, and individual-student-within-classroom differences.

Dispersed educational decision-making appears to be related both to differences among schools and to differences among classrooms within tracks in those schools. Certainly broad social differences (socio-economic status as indicated by housing type, etc.) may relate to sources of variance in these scores, but they seem likely to be mediated through access to different curricular opportunities which are shaped by the policies of independent education systems. Differences among classrooms within tracks seem likely related to differences among teachers in their decisions about curriculum, their subject matter knowledge, and their instructional practices. Differences among tracks would seem to be attributable to a combination of decisions deliberately seeking to create different curricular profiles of educational possibilities, and teacher enactments of curricular decisions based on their expectations of the track with which they are working.

We earlier explored 20 specific topic areas in eighth grade mathematics. We can estimate the variance proportions attributable to school, track, and classroom differences separately for each of these topic areas, rather than for the unscaled total score. These analyses provide further evidence that dispersed decision making and tracking exacerbate, if not create, large classroom and school variance components.

Combined variance components associated with classroom, tracking, and school differences range from about 25 percent for the two areas of 'whole numbers' and 'polygons and circles' (hence individual student differences were much larger in these areas) to as much as 73 percent for 'equations and formulas.' In such a case, individual student effort is more limited in what it can accomplish, a fact unsurprising since this content area is one that differs greatly in its coverage and emphasis according to which track a student is placed (see Chapter 2).

For 'equations and formulas', most of that variance is between classrooms and tracks. The component for differences across schools is larger when we restrict the sample to schools in which both tested classes were algebra (about 50 percent for school differences). This suggests that there is a more common notion of the content of algebra within schools than among different schools. Most other topics are widely taught as a part of all tracks. As a result, the classroom component of variance is much smaller for these other areas (about 25 to 30 percent with the sole exception of 'common fractions' which has about 45 percent).

By eighth grade, most of the achievement variation is 'natural' in other content areas due to differences among students and far less to differences related to systemic factors. For example, whole number arithmetic is one such topic with little variance attributable to issues of access and school differences. Still other areas such as fractions lie somewhere in between.

CONCLUDING REMARKS: ACCESS TO CURRICULUM MATTERS

This chapter has reported a number of analyses related to examining achievement variation in terms of systemic sources such as differences among schools, tracks, or classrooms within schools. We discussed whether these sources were related to characteristic systemic features of US education, such as dispersed decision-making and differential access (especially as tracking in mathematics). A number of conclusions seem to follow from these analyses. Among them are the following:

- The US population is heterogeneous culturally, ethnically, and racially. However, US achievement variability is not that noteworthy against the background of comparable variability in other TIMSS countries. This is especially true when the systematic relationship between variability and average achievement is taken into account.

- Unfortunately, our best students are not among the best from all TIMSS countries, especially for thirteen-year-olds in mathematics. Our best are not 'world class' in mathematics at either level, nor in science for thirteen-year-olds.

- The likely correlates of variability are our tracking practices in mathematics and our layered specialization approach in science (from at least the middle grades on).

- Achievement variability is a naturally occurring phenomenon reflecting variation among persons. However, patterns in achievement variability suggest that in mathematics education it is not simply a matter of the natural differences among persons. Much achievement variability seems related to structural factors.

- US achievement variability for nine- and thirteen-year-olds in mathematics was not so much *natural* as *created*. It appears to have been influenced by the structure of our educational systems, especially by our characteristic decentralized decision-making and by our practice of tracking.

Many would like to defend our characteristic practices of dispersed decision making and tracking since they are so deeply a part of the American approach to education. As tempting as it may be to defend them, they appear likely to be practices that limit individual students' opportunities to enhance their achievements. The unsatisfactory achievement levels for nine- and thirteen-year-olds call into question justifying these practices on the basis of their efficacy.

Exhibit 6.7. *Sources of Variance for 20 Eighth Grade Mathematics Scales in the US.*

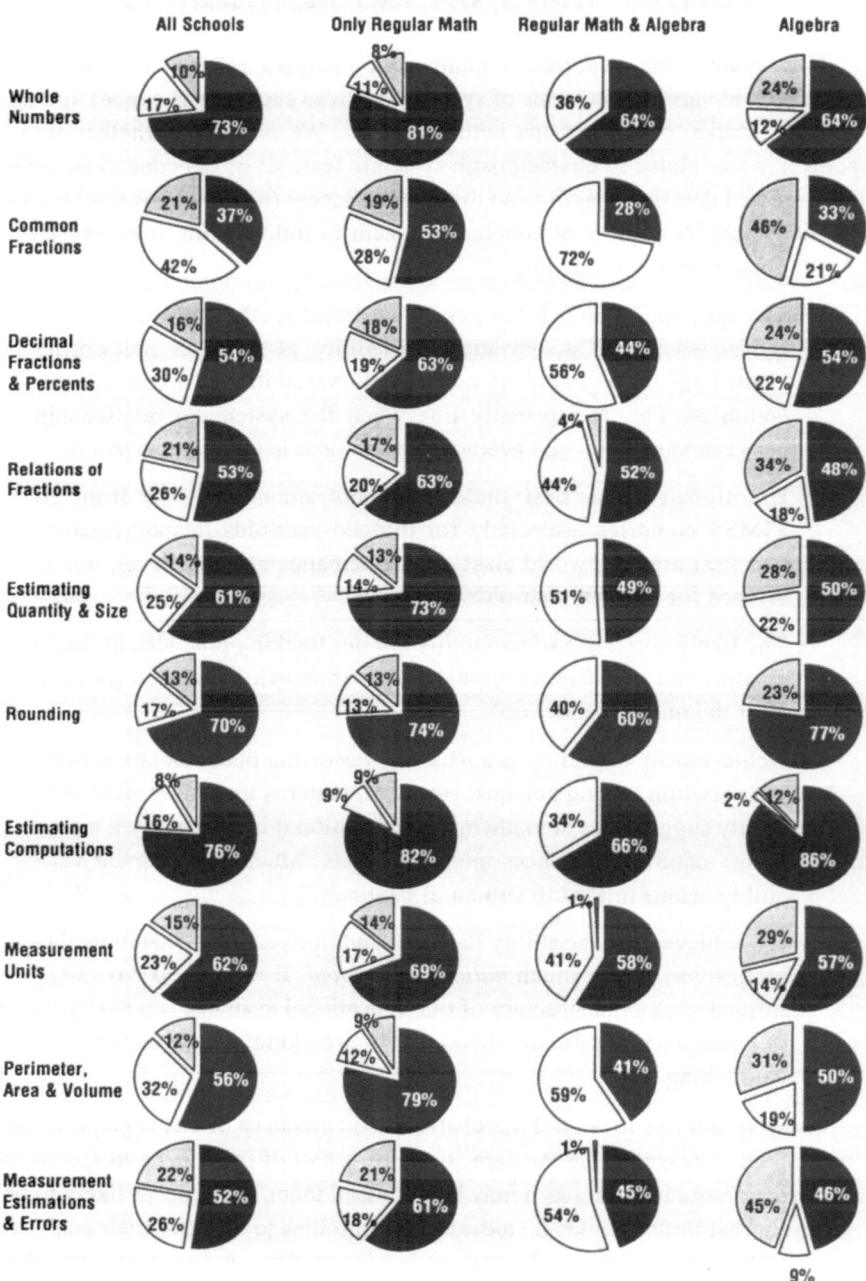

Exhibit 6.7. *(Continued).*

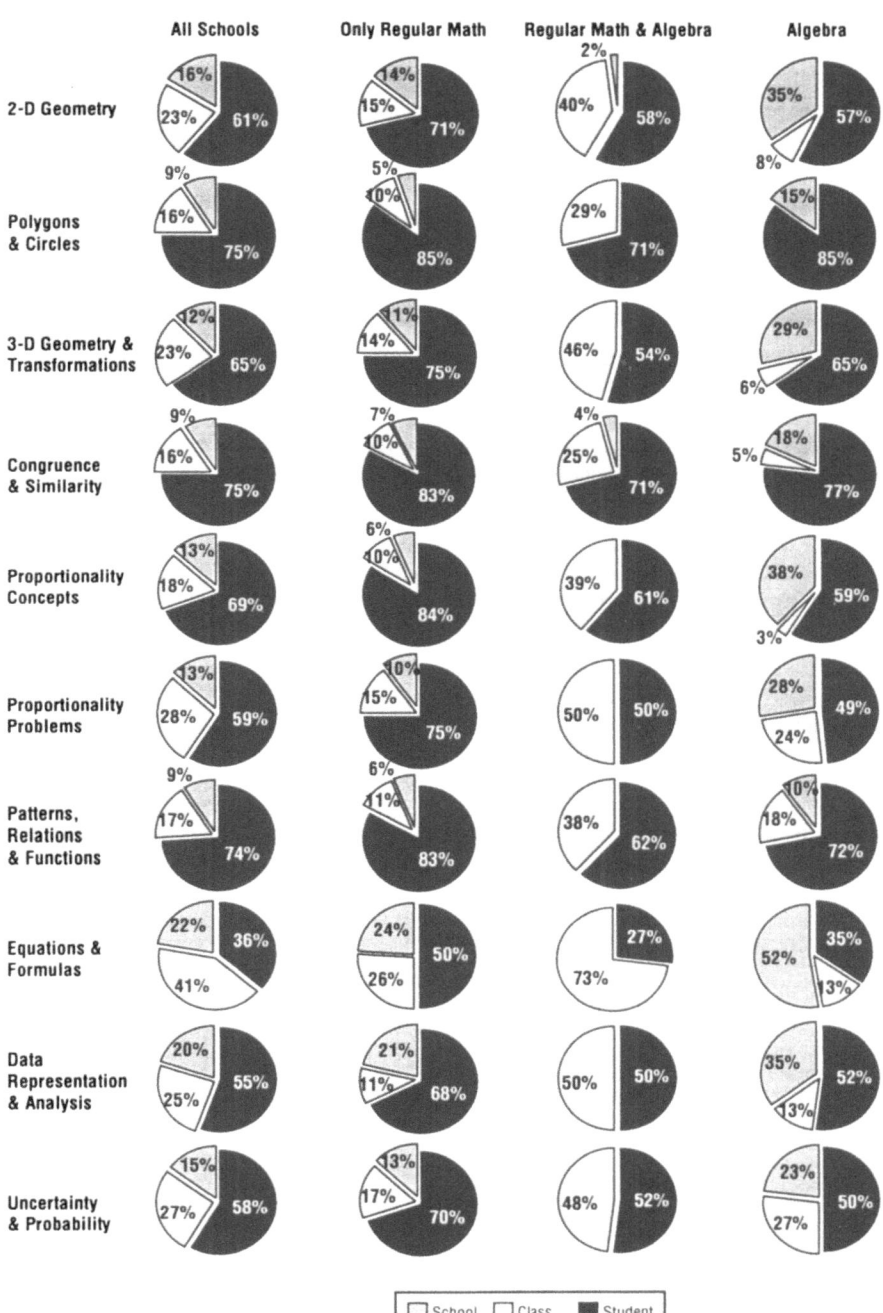

The analyses of this chapter suggest further that access to curriculum matters and that US practices distribute that access in a way different from most other TIMSS countries, and in ways that limit the efficacy of individual student efforts. Given the unsatisfactory achievements echoed in the twelfth grade results (see the last chapter below), surely it is time to stop, question these traditional practices, face some of their potential undesirable side effects, and turn to a search for new solutions to 'even the playing field' for all US students.

Notes–

[1]   We used resistant regression methods to determine the line so that it would not be appreciably affected by a few unusual values.

[2]   The global mathematics and science scores for the US and other TIMSS countries reported here and elsewhere were based on scaled combinations of items. However, as previously discussed, such scaling tends to deliberately focus on commonalities among disparate items and take a narrower view of what is estimated as common mathematics or science achievement. This has consequences for estimates of variance because the pool of variance and its sources differs from that for unscaled combinations of similar items.

[3]   'Imputed' in this context means that we must estimate likely scores of some items for each student on the basis of their other responses and subject matter category scores since not every student took every item in the TIMSS design. To form total scores for each student so that we can estimate the variance in those scores over the pool of students requires that we have at least an estimated response for each student on each item. That estimate does not have to be a 1 (correct) or a 0 (incorrect) but can be a decimal between 1 and 0 which represents the probability that the student would have answered the item correctly based on their responses to related items.

Part III
Hypotheses, Hunches, and Conclusions

In Parts I and II of this report, we presented details of some empirical findings from TIMSS. We looked at what mathematics and science were taught in US schools, how they were taught, who was taught what content, how schools functioned, what US teachers and students believed, and how curriculum and differing access to curriculum appeared to be related to educational achievements and gains. Now we turn to another sort of analysis in an attempt to answer the question: Is there a main story underlying all of these empirical stories about US science and mathematics education?

In this new analysis, the authors admit frankly to moving beyond the empirical data to speculate about relationships that underlie those data. The data in Parts I and II required simple description and low-level inferences. They described interesting – and, we believe, important – phenomena about US science and mathematics education. However, those descriptions dealt with the details, the pieces, of educational practice. An appropriate goal is surely the combining of these described pieces into a reasonable model of how they are related in a system that shapes and distributes access to educational possibilities in studying mathematics and the sciences. Such a model should be a portrait not just of static relationships, but of the dynamics of how educational practice functions, especially in the US. Only by understanding these relationships and dynamics more clearly can effective policies be shaped by empirical information.

Unfortunately, our work is not at the stage where these underlying relationships and dynamics can be clearly delineated empirically. On the other hand, through almost ten years of hard work, interacting with scholars from other countries, doing observations in seven countries and compiling several volumes of reports about TIMSS, the authors feel that even were we to speculate, those speculations would be informed by the data we have examined so long and its context in other research. We believe that we can make some guesses and form some hypotheses for further study that at least conceptually integrate multiple sources of information. While some of the things we discuss below are hunches – our best guesses based on a sense of the data – one central conjecture is a little more certain. It is a more global hypothesis that links a larger variety of empirical data. We believe it to be a hypothesis worthy of further investigation, although we offer it not as an empirical conclusion but only as a possibility for study and discussion.

Why should we as authors try to make this leap from describing to hypothesizing? We put our hypothesis forward because it seems to be a story that needs to be told, either to be supported or proven wrong by further examination with the TIMSS data and work beyond TIMSS. If the hypothesis proves correct, it offers a revealing picture of how the US' many educational systems work (or fail to work) to produce the kinds of educational results we see in US science and mathematics education. In addition, the systemic structures seen more clearly should have a fruitful impact on educational change. We believe serious reform and change is needed in US mathematics and science education. We also believe that such reform can only happen if we understand the TIMSS results in a broad, systemic context. Our formal analyses cannot yet take us to these levels. However, the need is great for the US national debate about these matters to be appropriately informed by empirical results and, if possible, accelerated towards improved educational policies. We offer what follows in the interest of stimulating informed debate and that of fostering a faster, but thoughtful, search for more conclusive global findings.

We undertake this task in several ways. In the first part of Chapter 7, we put forward our central hypothesis about systemic characteristics and how they relate to mathematics and science educational practice as documented through the TIMSS data. We examine what we believe to be among the more fundamental consequences of the accretion of American educational choices and beliefs: structural characteristics of educational practice in America. While this is clearly not the only possible explanation for what is observed in the TIMSS data, these systemic characteristics seem linked to educational practice in important ways. We believe them worth exploring for the light they shed on what is likely to work and not to work in efforts to change US science and mathematics education.

In the second part of Chapter 7, we offer some suggestive descriptions of possible ways these structural characteristics might be linked to empirical findings about science and mathematics educational practice as revealed by the TIMSS data. We summarize the results and integrate results from this and previously released TIMSS reports. We discuss some of the more visible features of US science and mathematics education. We hope both to organize some of the implications of the TIMSS data and to suggest an underlying story about fundamental consequences of deeper, but less visible, educational choices in America.

If there is an underlying story, what does it tell us about what we must do to secure a better educational future for American children learning mathematics and science? In Chapter 8, we suggest that there is not likely one single factor, even curriculum, that makes all the difference. We discuss the unsuccessful search for monolithic solutions to our systemic problems in mathematics and science education. We also discuss what this suggests about the likelihood of finding such a 'quick fix' in the future. The discussion in this chapter is less conjectural but it does extend the range of implications that seem to follow from the data presented.

In Chapter 9, we move beyond simple description in yet another way. The authors believe that the TIMSS data, carefully analyzed, *do* tell important stories about current US mathematics and science curricula and instruction. In the first part of this chapter, we try to summarize what we believe we have learned from a closer look at the TIMSS results. We briefly summarize some of the main lessons – the cross-cutting themes or stories – that we believe we have learned by looking at the TIMSS results through the eighth grade. In the second part of this chapter, we venture beyond these results to briefly consider future research that might usefully follow up on what we have learned from TIMSS.

Near the end of this report's preparations, the TIMSS results were released for students in the final year of secondary school. Formal analyses linking these results to those for earlier grades remain for a later report. The authors feel, however, that they would be remiss if they did not briefly summarize these results and offer some suggestions about their possible relationship to those for earlier grades. Chapter 10 concludes this report by offering just such a brief discussion.

Chapter 7
Systemic Features, Following Consequences: A Hypothesis

The authors believe that American education is shaped by an accretion of choices. Some are made directly by the people. Voters choose who will represent the community's educational interests through local school board elections. Voters elect political leaders – governors, legislators, the chief state school officer or state superintendent of education (in some states), members of the US Congress, and even the US president. All of these elected leaders have an impact on educational policies. Communities also vote directly on school issues and funding. Less directly, American citizens shape the climate for education with their attitudes towards education itself and towards schools, teachers, and who should be responsible for them. They also do not hesitate to express their societal values, preferences, opinions, and whims – all faithfully documented by polls and surveys, and colorfully reported by US mass media.

Elected leaders respond to these expressions. Resulting educational policy is a combination of the agendas of policy makers, political leaders (who attempt to represent their constituencies and respond to what they perceive as their constituencies' preferences), and special interest groups at all levels – special interest groups with their own agendas for educational matters.

More specific choices are made by professionals paid to accept responsibilities for detailed decision making on a daily basis. Neither the American citizenry directly nor the elected representatives can take on-going responsibility for the stream of decisions required to keep education systems and schools functioning on a daily basis. Citizens and their elected representatives can give broad shape and bounds to these decisions through their influence on educational policies. However, detailed policies and decisions must be delegated to those who have operational responsibilities and oversight for schools, local school districts, state education systems, etc., in which policies and decisions are implemented.

We believe that the accretion of those choices and decisions – and how they are shaped by our beliefs, values, and preferences – have created current US mathematics and science education. No single choice could have shaped our current status. Because they are the result of many choices over time, our educational systems lack the internal coherence found in systems more deliberately designed. Trends in educational choices have ebbed and flowed over the years, reflecting perennial and ever-changing concerns of the American people, responding to transient issues that capture the public's attention for a time, and expressing the agendas of groups that have particular viewpoints about education.

US mathematics and science education, as have many other components of US educational practice, have reflected those changing trends. Science and mathematics education have moved from traditional approaches to 'new' mathematics and science and back again to 'basics' that claimed to restore traditional methods and values. They have responded to various calls for reform and counter-reform and, as usual, are currently locked in the struggle between deliberate change versus preserving a status – a status thought by many to be sufficiently satisfactory to eliminate the need for radical change.

Even a single choice usually has many consequences. The wide river of continued choice we have just described certainly does. Surprisingly, the choices made by many persons in many places have not resulted in educational practice completely without form or feature. Certain choices, stemming from Americans' deeply held beliefs, have been widely repeated over the years. Some of these choices seem central to the nature of US educational practice and US education systems. They seem likely to have consequences sufficiently fundamental to have helped form the characteristic shape of American mathematics and science education. We are concerned primarily with three such systemic features and the impact they have had on how educational possibilities are provided for mastering mathematics and the sciences in US elementary and secondary school.

THREE STRUCTURAL CHARACTERISTICS OF US EDUCATIONAL PRACTICE

While many structural characteristics are undoubtedly implicated in the present shape of American science and mathematics education, we limit ourselves here to examining three. We believe these have been shaped by the accumulation of our educational choices and are particularly important in shaping US mathematics and science education. A more detailed accounting of the evolution of these systemic features we leave to scholars of the history and politics of educational practice. We hold that the features of US educational systems are a matter of fact. How they have come to their present state is beyond the hypothesis we seek to set out here.

DISTRIBUTED DECISION MAKING

The United States from its beginnings has struggled to find the balance between being 'one nation' and a federation of 'united' but independent states. Before there was the Constitution for the 'United States of America,' there were Articles of Confederation for a looser federal union of independent states. One source described what 'united' the States in 1787 at the birth of the Constitution as essentially "a mutual treaty conveniently dishonored on all sides."[1]

The struggle to balance the general (common) welfare with the freedoms and rights of individual states has never abated. Today it is again a focus in public debate. Suggestions have varied from 're-inventing government' to devolution, the latter an approach that has been characterized as "the rearrangement (rather than reform or diminution) of public authority."[2] Devolution is a strategy based on the tenth article of the Bill of Rights, stressing that all powers not explicitly delegated to the US federal government 'devolve' back to the states and, echoing the language of the tenth amendment, 'to the people.'[3] In today's debate, devolution not only implies shifting power from the federal government to the states but 'to the people' in local communities as well.

Devolution has implications for education.[4] US tradition in public education has long reflected shared responsibility for educational goals and policies – responsibilities shared by local school districts, states, and, to a lesser extent, federal agencies.[5] Governors and states do not have a constitutional responsibility for education but, as one author wrote, "Decentralized education policy is not enshrined in the Constitution.... It is rather a matter of tradition, political preference, and the accumulation over time of institutional capacity."[6]

One practical result of this tradition of shared responsibility and 'decentralized education policy' is distributed educational decision-making. Decision-making is distributed among the many levels, agencies, and social actors responsible for official educational choices in the US and its communities. We have traditionally distributed educational responsibilities not just to states, local communities, and school districts, but even to individual schools and teachers.

Unfortunately, 'distributed' in this context is a deceptive term. When decision-making is dispersed to less global sites, the conditions at those sites strongly affect the decisions made. In general, decisions are made within the 'decision horizons' – the limited information and criteria – of the decision-makers. When educational decision-making is dispersed, decisions are made with more limited decision horizons. They are based on the information locally available and on local goals (rather than on broader or more aggregate goals such as 'creating a mathematically and scientifically literate society'). This is both a source of strength and of weakness.

Were decision-making simply a matter of the decisions made at one local site, it would be less problematic. With multiple local sites for important educational decisions – each site with its own goals and information – there is a potential for conflicting, even competing, decisions.[7] As a people, we have long held that competition breeds excellence. However, widespread competition in performing similar services can breed redundancy, cautiousness, sensitivity to public image, unthoughtful inclusiveness, a lack of strategic focus, and other concerns.

We are not necessarily suggesting as a part of the underlying story of the TIMSS results already discussed that shared responsibility is bad. More simply, we are suggesting that educational decision-making *is* decentralized (for better or worse) and that this is a characteristic, structural, and systemic feature of US education. Further, a case could be made that this has consequences that directly affect the quality of science and mathematics education.

The TIMSS data were used to examine the use of shared responsibility in decision-making in the US and other TIMSS countries by looking at decisions about science and mathematics educational goals, course offerings, course content (syllabi), and textbooks. Just over 40 percent of all the decisions were made solely at a central site (an education ministry or regional authority). Almost 17 percent of these decisions were made solely by teachers and another six percent solely by processes at individual schools. Just over 10 percent of the decisions were made solely by other decision-making individuals or agencies.

From these data we can see that, among the 30 countries, most decisions were made at a single level. Most decisions were, in fact, made at a central location. Only about a fourth were joint decisions. Among TIMSS countries, there was little shared responsibility in these decisions about mathematics and science education matters.

There were some interesting variations from this general pattern. Although approximately 25 percent joint decisions were made for educational goals and course content, about 35 percent of decisions about course offerings were joint or shared decisions. The rest were made almost entirely by central authorities. Over 42 percent of textbook selection decisions were shared decisions. This was the largest exception to the 'sole decision-maker' pattern.

Decisions about content in course syllabi were made jointly in about 22 percent of the cases. For 62 percent of the cases in 30 countries, those decisions were made at a ministry or other central authority. Only about five percent of these decisions were made with teachers as the sole determiners of course content.

Unsurprisingly, about 80 percent of decisions on general goals for an education system were made by central authorities. The remainder were made mostly as joint decisions which included central authorities among the decision makers. In fact, if a single 'decision-making actor' made any decision on content goals, even more specific goals, that 'actor' would be a ministry or similar central authority in 95 percent of the cases. Teachers had almost no official role in goal setting. Overwhelmingly across the 30 countries, the input provided by teachers was, at most, advisory.

Thus, in these 30 countries, joint decision-making was rare, especially in decisions about the goals and content of instruction. These decisions most often were made solely, or in some cases jointly with others, by a central ministry or regional authority. Teacher input was typically advisory only. This picture contrasts strongly with the pattern for the US in which no such decisions were made centrally and where teachers often act as determiners of course content either by design or default.

These data suggest that not only was distributed decision-making a characteristic feature of US mathematics and science, but that it was almost uniquely so among TIMSS countries. While there was some distributed decision-making in other TIMSS countries, and about a fourth of the decisions were joint, most decision-making was concentrated in the hands of central authorities. We do not take this to imply that such centralization is good, needed, or even more effective than decentralized decision-making. Those are either empirical questions or matters of opinion. We simply point out that it is a structural characteristic of US education different from what was common in other (TIMSS) countries.

DIFFERENTIAL ACCESS TO EDUCATIONAL CONTENT

The data in Chapters 2 and 6 show that differential access to educational possibilities for different groups of students has been a common practice in the United States. For example, tracking within schools, either as official policy or implicit practice, exposes different children to different mathematics curricula. Even done in the name of educational efficiency (by providing teachers with classes having similar mathematics abilities and preparation), the result is that different children receive instruction using different textbooks covering varying topics to which unequal amounts of time are devoted – even within the same school. Data already presented (in Chapter 2) make this clear.

The data seem to imply that differential access to educational possibilities is a structural, systemic characteristic of US educational systems. This is in marked contrast to most other countries (as discussed in Chapter 2). Given this broader context, it appears that differentiation in the form of tracking is more a US choice and strategy rather than an essential feature of educational systems. We have already discussed in Chapter 6 how such differential access is related to the educational achievements of American students.

THE MIDDLE GRADES: ENDING ELEMENTARY SCHOOL
OR BEGINNING SECONDARY?

A third structural feature contributing to the current state of US science and mathematics education is with which other grades US schools group the middle school grades. The phrases 'middle grades' or 'middle school' perhaps have come into usage because of some ambiguity as to what part of education these years belong. There is great variance among nations in whether the middle grades are grouped with secondary school grades or with elementary school grades. The latter is the case most often in the US, although this is often hidden by the terms 'middle grades' or 'middle school.'

We, the authors, believe that we can make a strong case that the middle grades – in particular, seventh and eighth grades – are curricularly and substantively considered the culmination of elementary school, rather than the beginning of secondary school, in the US. There is considerable evidence to support this contention. We have earlier discussed how the grades were organized in the US and several other TIMSS countries. For the US, there were only a few cases in which at least some grades below seventh were not included in the same schools that contained seventh and eighth grades. Equally striking, there was only a small percentage of US schools in which a grade higher than ninth was included in the same school with seventh and eighth. This was in marked contrast to many other TIMSS countries.

This structural feature may, at first, seem without serious consequence. That seems clearly not to be so in the US. Uncertainty about how to provide for the middle grades and the common practice of considering them as the culmination of elementary school seems frequently to have had impacts on who taught students in those grades, the textbooks that were used, what topics were taught, and the way time was distributed among those subject matter topics. It resulted in students taught by individuals less prepared in mathematics and the sciences, the use of textbooks more like the general texts of elementary school than the more specialized texts of secondary school, and a continuation of the common lower grade strategy of devoting small amounts of time to many topics. Since in the middle grades most topics already introduced in lower grades persisted, curricula were often weighted towards review and only marginally extended the content already covered. Few new topics were introduced during the middle school years. Far from harmless, this practice appears to have helped make middle school mathematics and science content less rigorous and very different from that of the majority of TIMSS countries.[8]

Grouping eighth grade with the end of elementary school rather than the beginning of secondary is not simply a matter of grade structure alone. It is also a matter of academic substance. Other, high achieving countries (the Czech Republic, Korea) also occasionally group eighth grade with the end of elementary school. However, in those countries, an analysis of their curricula reveal that curricularly, if not structurally, the content of mathematics and science instruction in eighth grade is the beginning of secondary school content rather than the culmination (or repetition) of elementary school subject matter. That is not true in the US. Perhaps what is really at work in the US is the lack of a clear, strategic curricular vision of what is to be accomplished in each grade – a lack of vision not shared by the Czech Republic or Korea.

This chapter focuses on the authors' hypothesis that there is a story behind the TIMSS data about what is taught, to whom it is taught, how it is taught, and what is achieved. So far, we have put forward a set of three systemic features we feel to be broadly characteristic of American mathematics and science education practice. We believe each of these reflects national beliefs and values, and is an accretion of many choices. We also believe each has serious consequences, consequences we have already seen reflected in the empirical data presented earlier in this volume and in TIMSS data reported elsewhere.

In the next part of this chapter, we turn to a survey of the state of US mathematics and science education as revealed in the TIMSS data. We believe that this status is related to the systemic characteristics we have just outlined, but that is a decision each reader must make for herself or himself. If there is a connection, then there is an underlying story. That, at least, is the authors' hypothesis.

The story is that many of the less-than-desirable features of US science and mathematics education described in the rest of this chapter are linked to the three previously discussed systemic characteristics of American education. They are separate symptoms rather than separate diseases. If this underlying story is true, the implications are profound. Certainly it implies that we must seek the disease rather than the symptoms and that any remedy must be systemic rather than piece-meal and palliative. Surely this possible linkage warrants serious consideration.

THE CONSEQUENCES THAT FOLLOW: OUR HYPOTHESIS CONSIDERED FURTHER

We believe that the TIMSS data reported in Chapters 2 through 6 provide strong empirical support for seven features of US science and mathematics education that were true at the time those data were gathered and likely continue to be true. We also believe that the story of where we stand summarized in these features is strikingly disappointing. Further, we believe that these seven features are likely related to the three systemic characteristics we have presented thus far in this chapter. This not just a story about US students' largely average comparative achievement in science and mathematics. It is about the more visible surface features of the story of how American educational choices have evolved to produce the current state of US science and mathematics education, at least in part through systemic weaknesses that would need to be addressed for any large scale and lasting improvements.

For the rest of this chapter, we review and summarize empirical evidence for seven key features of US mathematics and science education. Only things that are understood can be changed. Here we try to further our understanding of the US mathematics and science education *status quo* by focusing on seven of its features. We will have to re-state some key findings from *A Splintered Vision*, findings that used analyses of curricula, textbooks, and questionnaires. We will integrate them with analyses reported in Chapters 2 through 6 of this report.

A SPLINTERED VISION OF EDUCATIONAL GOALS

The TIMSS data strongly suggest that US mathematics and science education practices reflected a splintered vision of what should and could be done. This was quite clearly true at the time the TIMSS data were collected from1993 to 1995 and likely continues even now. This disarray seems to have resulted from the lack of broad, intellectually coherent, commonly accepted, guiding visions of education in mathematics and the sciences.

As pointed out in *A Splintered Vision*, we "have no shortage of visions of how US mathematics and science education are 'supposed to be.' Partly this is because we have no shortage of sites articulating visions. In the US, shareholders in the 'official' vision enterprise include more than 15,000 local school districts and boards, 50 state education agencies, and various federal offices, committees, boards, and administrators." These 'shareholders' were suggested as including: "textbook and test producers, members of professional organizations in mathematics and the sciences, teachers' organizations, special interest groups with educational goals, government officials at all levels who state policy broadly," and so on.[9]

These visions are not filtered through any formal or central consensus-seeking mechanism or agency. They are acted on, at a minimum, by the 50 state education agencies and likely by most, if not all, of the more than 15,000 local school districts and boards. Thus, we have policy 'polyverse' in mathematics and science education (as well as other areas) since we have education systems, not an education system. Not all had (or likely have now) the same visions or the same 'take' on goals for education in mathematics and the sciences. They express their visions and goals in the policies they set, the curricula they put forward, the selection of textbooks which they influence and control, and the monitoring and distribution of resources to teachers and classrooms. Evidence for this 'splintered vision' must be seen in data for these expressions – for official curricula, adopted textbooks, and teachers' practices. They are examined in several of the features reviewed below. Those data forced us to conclude that, to the extent that we might even legitimately speak of a US 'national' vision for education in the sciences and mathematics, that vision was splintered and shared only in the broadest terms.

OFFICIAL CURRICULA THAT ARE 'A MILE WIDE AND AN INCH DEEP'

As we stated in Chapter 2, we found official US mathematics and science curricula typically to have been 'a mile wide and an inch deep.' Those curricula reflected only broad common goals for school mathematics and science among states and districts. Certain features occurred so frequently that they characterized 'national' composite curricula – an intersection of commonalties found in at least 70 percent of a representative sample of state curricula.

That composite showed typical curricula to include far more topics than other countries. Exhibits 1 and 2 of *A Splintered Vision* showed that, in the composite, the US typically covered far more mathematics and science topics at every grade than most other TIMSS countries.[10] We did this in mathematics through eighth grade, falling below the seventy-fifth percentile only in the ninth through twelfth grades, and even then staying well above the median number of mathematics topics covered in other TIMSS countries. In science, this was true through the eighth grade. In tenth through twelfth grades, US science courses were sufficiently specialized to place the number of topics covered below the median for TIMSS countries (tenth grade) and even below the twenty-fifth percentile (eleventh and twelfth grades).

Exhibits 11 through 13 of *A Splintered Vision* showed that in both fourth and eighth grade mathematics and fourth grade science the topics that survived in composites – that is, topics we had broadly in common across most state education systems – were still more diverse than the corresponding composite of other TIMSS countries.[11] The same was not true in eighth grade science. This does not imply, however, that each state's eighth grade science curriculum did not have more topics or more diversity, although that may have been true. It might instead imply that there were fewer commonalities among these science curricula across states, and thus fewer topics entered the composite. The additional diversity from local district variance was not even considered in these data.

The data also showed that, in the composites, topics persisted far longer than cross-national averages, that we added more topics and dropped fewer, and as a result we typically continued to devote instructional attention to topics far longer than most other countries. Since mathematics and science instructional time was limited, most topics were necessarily covered only briefly, a fact directly supported by the TIMSS data on US textbooks, as was the fact that this coverage was largely without serious demands on students.

These official curricula expressed intentions. They had their reflections in US textbooks that reacted to composite curricula (obviously weighted towards more populous and influential states) and in the practice of teachers who had responsibilities for translating those intended distributions of educational possibilities into classroom realities. Thus, further supporting data can be seen in those for textbooks (reviewed below) and in characteristic teacher instructional practices (with what we taught, examined in Chapter 2 here).

Taken together, this evidence seems sufficient to conclude both that the US typically divided time among many topics both in mathematics and in science, that it necessarily intended that only small amounts of instructional time be devoted to most of those topics, and thus that each topic was covered in a comparatively simple form without challenging the students to complex or demanding performances. This seems clearly to justify labeling US official science and mathematics curricula as 'a mile wide and an inch deep.'

CAUTIOUS, INCLUSIVE TEXTBOOKS

Textbooks and tests helped shape implementation, but further reflected fragmentation rather than adding coherence. At each grade, with few exceptions, US science and mathematics textbooks provided support for instruction on a large number of topics, many of which had been covered in previous grades, but for which the textbooks continued to provide extensive review material.

Composites were again created for topics that appeared in 70 percent of a representative sample of US textbooks. Separate composites were made for mathematics and for science in fourth and eighth grade textbooks, and in textbooks for 'specialist' (calculus and physics) courses typically taught at twelfth grade. Exhibits 19 and 20 in *A Splintered Vision* showed that those composites placed the US far above the seventy-fifth percentile of TIMSS countries in the number of mathematics and science topics at all three grade levels.[12]

Exhibits 21 through 24 displayed data on the five topics to which the most space was devoted in the composite of US textbooks in both fourth and eighth grade science and mathematics.[13] With the exception of eighth grade algebra books and eighth grade books for specialized, single-discipline-oriented science courses, the total amount of space devoted to the five 'most covered' topics accounted for from just over 25 percent to just over 50 percent of the space in textbooks for both school subjects at both grade levels. In all cases, this was far less than was accounted for by the five 'most covered' topics in contrast nations (Japan and Germany).[14]

Other data displayed in *A Splintered Vision* showed that eighth grade science textbooks focused primarily on student knowledge, recognition, and recall of subject matter facts and routine procedures rather than on problem solving, analyzing, theorizing, or investigating.[15] The United States differed little from the other TIMSS countries in doing so. Eighth grade mathematics textbooks focused primarily on recalling and recognizing mathematical objects (their names and their properties), on using routine procedures, and, to a much smaller extent, on solving problems. This was in contrast to material on reasoning, justification, generalization, and communication of mathematical results and ideas.

These data certainly justify characterizing the content of US science and mathematics textbooks (and, presumably, the 'official' curricula that were market forces driving the development of these textbooks) as undemanding. Students may find it challenging to master a large body of facts and skills with little time devoted to each. However, this is far different from curricula, textbooks, and teachers that expect students to be able to perform complex, integrated tasks with the content they are learning (plan investigations, critique arguments, etc.). It is in this sense that these data show US science and mathematics textbooks to have been undemanding.

These mathematics and science textbook series had books that were large, inclusive, and filled with redundant content over the grades.[16] These textbooks made it possible for teachers to select from many possible topics but offered little guidance on how or why to choose some topics, emphasize some, and altogether omit others. Certainly the data justify characterizing the representative US textbooks examined as 'inclusive.' The focus on comparatively undemanding performances by students seems a cautious approach, leaving more demanding tasks to be supplied largely by teachers as supplements to more routine textbook material.

While it must remain conjectural, these books seem to reflect publishers responses to a textbook marketplace in which no strong, shared visions and goals shaped the content to be included. This seems further to justify characterizing these textbooks as cautious.

CLASSROOM INSTRUCTION LACKING COHERENCE

We can summarize the findings from the TIMSS data on US mathematics and science classroom instruction by saying that that instruction lacked coherence. US eighth grade science and mathematics teachers virtually never devoted a large number of periods to any one topic, as can be seen in Exhibits 29 and 30 of *A Splintered Vision*.[17] Even examining the time devoted to the five most emphasized topics showed that US teachers devoted surprisingly little time to those topics compared to other TIMSS countries such as Japan, especially in general mathematics and science courses.[18]

Chapter 2 of this volume showed that the number of topics to which US fourth and eighth grade science and mathematics teachers devoted their instructional time was large at both grade levels. It further showed the small amount of time allocated to each of those topics, with one or two exceptions. A comparison of achievement results (the gains) in specific content areas (see Chapter 5) further suggests that achievement gains in specific content areas were likely related to instructional time devoted to them. These patterns of gains seem to be captured also by the phrase 'inch deep' as the US was the only nation not to be in the top tier in at least some area of the curriculum.

Beyond this splitting of attention among large numbers of topics, US fourth and eighth grade mathematics and science teachers also reported instructional patterns in typical lessons that moved among large numbers of activities. US teachers conducted far more activities per period than was typical of teachers in many other TIMSS countries. They appeared mainly to orchestrate large numbers of activities, and to have become skilled at moving among them to secure attention, providing on-the-spot review, presenting instruction on some new content, and allowing in-class time for students to become 'engaged' with homework tasks.

Other related TIMSS results support the idea that mathematics classroom instruction was fragmented and lacked coherence.[19] Mathematicians were asked to rate the quality of the mathematics presented in videotaped mathematics lessons from classrooms in the US, Germany, and Japan. The criterion for high quality was the degree of connectedness that the mathematical ideas had within a single lesson. These ratings revealed that not a single US lesson was rated as of high quality and high coherence while about 25 to 30 percent of Japanese and German lessons viewed received high ratings. Almost 90 percent of the US lessons were rated as low in quality and connectedness.[20]

HIGHLY REPETITIVE INSTRUCTION OVER THE GRADES

The TIMSS data revealed US mathematics and science instruction to have been highly repetitive over the grades. Exhibits 3 and 4 of *A Splintered Vision* show how many topics were added in both science and mathematics and how few were dropped in grade after grade until specialization began, typically in eighth grade for science and ninth grade for mathematics.[21] Exhibits 5 through 8 show that science and mathematics topics, on average, persisted longer than in most other TIMSS countries.[22]

Data in Chapter 2 of the present volume show the large number of topics to which US fourth and eighth grade teachers devoted attention and that likely resulted, in part, from topics that were dealt with in some form grade after grade. In fact, the data in Chapter 2 also show the pronounced degree of overlap in the mathematics curriculum between pairs of adjacent grades and even between fourth and eighth grades. This was less true in science, although there was still a fair degree of overlap across the grades. Official mathematics and science curricula from US education systems, as we discussed earlier, built redundancy into the curricular structure and typically planned for each topic to be covered in several grades. US science and mathematics textbooks (certainly those examined in detail at fourth and eighth grades) were inclusive partly because they provided so much review material and retained so many topics already introduced in previous grades.

Chapter 2's data show that teachers devoted attention to a large number of topics. They also showed that this resulted partly from curricular redundancy unchallenged by inclusive textbooks. Was much of the instructional attention devoted to previously introduced topics largely review rather than systematic journeys through topics' contents which spread across several grades? Chapter 3 showed that US teachers devoted a far greater proportion of instructional time to review than was typical in many TIMSS countries. Part of this emphasis on review might be due to providing on-the-spot prerequisite knowledge for new content to be introduced. Even with this motivation, however, the result was still continued repetition of previously taught content and unusually high proportions of time devoted to review.

The constrained and valuable resource of science and mathematics instructional time was devoted to repeated topics, topics that persisted grade after grade. That further limited the amount of time allocated to any single topic and, particularly, to new topics. Limited time, in turn, seems likely to have led to unchallenging activities for students. Teachers devoted unusually high proportions of instructional time to review, which exacerbated the need for simplicity and the concomitant lack of challenge in each topic. US instruction focused on low demands calling for factual knowledge and routine procedures and rarely making more demanding expectations such as for reasoning and theorizing (see Chapter 3). Given this highly repetitive but undemanding instruction grade after grade, it seems hardly surprising that US eighth grade students reported finding mathematics and science easy but boring.

THE MIDDLE GRADES' LACK OF RIGOROUS CONTENT

Mathematics and science instruction in the middle grades was highly repetitive and progressed little over the demands of earlier grades. This was, in part, the point of the previous section. However, we raise here a different question, 'Is there any evidence that this repetitiveness was evenly distributed across all the grades through eighth grade?' There appears, instead, to be evidence to the contrary. The number of mathematics topics covered in the US at each grade reached its peak in eighth grade, with only slightly fewer topics covered in seventh and sixth grades.[23] The corresponding peak for science occurred in ninth grade.[24] Both peaks were preceded by a continued increase in the number of topics covered in successive grades (although there was a large jump rather than a small increment at fifth grade for science).

. Far more topics than was typical cross-nationally were added in mathematics at every grade through sixth grade, the same grade in which US curricula were reported to have dropped topics for the first time (that number of topics dropped becoming unusually large at ninth grade).[25] Science curricula added unusually large number of topics at first and fifth grades with only modest additions elsewhere, but they were not reported to drop topics for the first time until ninth grade.[26]

These data clearly imply that instructional time was devoted to more topics at seventh and eighth grades than in any other grade. This was true for both mathematics and science. Further, given the patterns of adding and dropping topics, there were fewer new and more old topics at seventh and eighth grades than in any other grades. Mathematics and science curricular redundancy peaked in seventh and eighth grades. US textbooks, as we have seen, reflected – rather than challenged – the official curricular redundancy.

These data suggest strongly that the characteristic redundancy of US mathematics and science curricula, textbooks, and instruction was particularly marked at seventh and eighth grades. New topics typical of secondary education in mathematics and the sciences must have been delayed given the modest numbers of new topics introduced in seventh and eighth grades. Far from evenly distributed, the ubiquitous redundancy of US mathematics and science curricula, textbooks, and instruction seems to have been present at all grades but concentrated at seventh and eighth. This is consistent with the conjecture that these two grades are characteristically viewed in US schools as being the end of elementary school rather than the beginning of secondary school.[27]

What would be empirical criteria for a 'lack of rigorous content?' Surely the criteria would include repeated exposure to the same content that had been seen many times before. They would also include treating each topic only briefly and selecting topics typically (by international standards) covered in earlier grades. They would undoubtedly include devoting significant amounts of class time each school day to review of content already covered, both in the current grade and in previous grades. Certainly it might include being surrounded by peers who found content easy but boring and many of whom regarded the content as unimportant.[28]

If those are the criteria for a 'lack of rigorous content,' then the TIMSS data support the claim that US seventh and eighth grade mathematics and science constituted such a lack of rigor. The effects of that lack of rigor were reflected in the disappointing achievement of US thirteen-year-olds on the TIMSS tests, a low level of accomplishment not reflected in the achievements of nine-year-olds, especially in science.

'SOME GET AND SOME DON'T'

The TIMSS data support the notion that some students had access to educational possibilities in mathematics and the sciences that were denied to others. The evidence presented in Chapter 2 supports this claim. The analyses of achievement results in Chapter 6 provide further support.

Different students had differing access – and lack of access – to educational content. This was true for US schools, often reflecting differences among states and communities. It was true among tracks within schools, especially in mathematics where tracking was direct (but also in science which often received the 'backwash' of mathematics tracking). It was even true of different classrooms within the same school even when this was not due to tracking.

What does it mean in concrete terms to have had differing access to educational possibilities? It could have meant being taught in courses with different official curricula, curricula that affected the content covered, emphasized, and omitted in those students' courses.[29] It could have meant being taught in classes using different textbooks — textbooks that differed not only in title but in content, goals, and in the kinds of curricula they were intended to support — further setting limits on students' abilities to achieve by their own efforts. It could also have meant being taught by teachers who emphasized different content, included more review, and shaped their expectations of what students could do based on the type of course in which students found themselves. Most of all, it could have meant a devastating combination of these things.

The reasons for this differentiation varied. They were not the same in mathematics and in science. In science, differential access, according to the TIMSS data on US science curricula, most often occurred because specialized courses (earth sciences, life sciences, etc.) were taught and because some students were in one type and others in another. In mathematics, widely believed to build 'incrementally' on instruction from previous grades, differences occurred more often from explicit tracking than from content specialization. As a result, some students received instruction from textbooks and in classes that emphasized topics fundamental to later mathematics study (for example, algebra and geometry instruction in middle school years). Others did not (for example, receiving instead instruction that still emphasized arithmetic in middle school years).

Did differential access lead to differing achievement or did differing achievement indicate that there were good reasons for tracking students and providing differential access to educational possibilities? The US data alone cannot answer this question. However, apart from the inherent inequities already discussed which call differential access into question, the higher levels of achievement in many TIMSS countries that were not characterized by differential access, question not the legitimacy but the *effectiveness* of this traditional American practice. In US science and mathematics education, the TIMSS evidence supports the claim that 'some get and some don't.' It does not support the claim that the 'some' who didn't couldn't have responded effectively. Are American students so different from those in other countries or is it just our perceptions of them and our choices about the proper way to 'do school' for them that differ?

CONCLUDING REMARKS

No one who reads the preceding discussions of seven features we suggest are characteristic of US science and mathematics education, and who agrees with even some of them, can fail to see the situation as serious. We as authors are not trying to indict US schools or to suggest that they are fatally flawed or beyond repair. However, taking the data seriously carries its own message, a message that, it seems to us, concerned Americans would be unwise not to examine further.

It is not that teachers, textbooks, or curricula could not be improved. They can. Instead, it is because we believe that these improvements would be small, incremental improvements that could change the disappointing *status quo* of US mathematics and science education only in small ways. The fundamental problems are systemic. The fundamental improvements must be systemic as well. We believe that this is the underlying story of the state of US science and mathematics education and the structural characteristics of American educational practice that are related to them.

Systemic problems are only difficult, not intractable. However, they can become intractable with only a few limiting conditions on possible solutions. If the structural characteristics of US educational practice – some essential and some escapable – lie at the heart of America's problems in science and mathematics attainments, then they must be addressed in any effort at significant change. Since they are related to our societal and educational beliefs, values, and choices, addressing them may call for a difficult re-examination.

If our hypothesis of a systemic nature of the underlying problems is valid, significant change can come only by facing the consequences of what American beliefs and choices have produced and by searching for new ways to embody those beliefs that avoid non-essential systemic characteristics that contribute to our difficulties. If we cannot or will not face the consequences of our societal choices – if we deny that there are problems, continue to base proposed solutions on ideological positions rather than open investigation and discussion, or seek escape in a 'return' to something no longer possible if it ever was – then we set in stone those few limits that make our educational problems unsolvable.

Notes–

[1]   Amar, A. R. (June 1987). Of sovereignty and federalism. *Yale Law Journal*, 96(7).

[2]   Donahue, J. D. 1997. *Disunited States*. New York: Basic Books, page 7.

[3]   Article X Of the Bill Of Rights: "The powers not delegated to the United States by the Constitution, nor prohibited by it to the States, are reserved to the States respectively, or to the people."

[4]   See Donahue, *Disunited States*, pages 151-158.

[5] "By long-standing custom the federal government maintains a subordinate role in American education." Donahue, *Disunited States*, page 151.

[6] Donahue, *Disunited States*, page 153.

[7] Some of us (Schmidt, McKnight, and Raizen) have discussed these matters in *A Splintered Vision: An Investigation of U.S. Science and Mathematics Education*. Dordrecht: Kluwer, 1997. In particular, these matters were discussed from the perspective of an 'organizational process model' set out by G.T. Allison. (See his *Essence of Decision: Explaining the Cuban Missile Crisis*. Glenview, IL: Scott, Foresman and Company; 1971.)

[8] See Chapter 2 as well as *A Splintered Vision* (already cited).

[9] *A Splintered Vision*, page 90.

[10] *A Splintered Vision*, pp. 15-16.

[11] Pp. 31-38.

[12] Page 55.

[13] Pp. 57-61.

[14] This finding was more generally true compared to other nations as can be seen in Schmidt, et. al. (1997), *Many Visions, Many Aims*, Volumes 1 and 2.

[15] Pp. 63-66.

[16] Physically, US textbooks were the largest among all TIMSS countries according to the data on page area and number of pages.

[17] Pp. 73-74.

[18] *A Splintered Vision*, pp. 76-77.

[19] See Stigler, et al. (in preparation). *The TIMSS Classroom Videotape Study*. Washington, DC: Department of Education. National Center for Education Statistics.

[20] Similar findings were noted in a small scale study observing classrooms in six countries as a part of instrument development efforts for TIMSS. See Schmidt et al. (1996). *Characterizing Pedagogical Flow* Dordrecht: Kluwer.

[21] Pp. 19-20.

[22] Pp. 23-27.

[23] Exhibit 1, page 15, of *A Splintered Vision*.

[24] Exhibit 2, page 16, of *A Splintered Vision*.

[25] *A Splintered Vision*, Exhibit 3, page 19.

[26] *A Splintered Vision*, Exhibit 4, page 21.

[27] See Chapter 5.

[28] See Chapter 5.

[29] See Chapter 5.

Chapter 8
There Are No Magic Bullets

In the complex arena of educational systems there is a fine line between the simple and the simplistic. Everyone prefers simple solutions. No one can afford simplistic solutions for which there are no sound empirical grounds supporting their efficacy in education and, here, particularly in US science and mathematics education.

Single explanatory factors are not likely sufficient and could be considered as a sole solution to achievement difficulties only on the basis of substantial empirical support. There seems to be a strong *prima facie* case against single factors. US mathematics and science education takes place in complex educational systems. The problems are most likely to be systemic. Systemic problems most likely require systemic solutions. Most of the weaknesses discussed in the preceding chapters not only concatenate with each other, they are mutually destabilizing and each likely exacerbates the effects of the others. For example, a lack of clear goals and focus perhaps could be overcome by a teaching force with strong subject matter knowledge. If, as the evidence suggests, US science and mathematics education suffers both from a lack of clear goals and a less-than-strongly knowledgeable teaching force, the combined effect of these two factors is likely to be multiplicative rather than simply cumulative.

If this reasoning is sound, it suggests that it is fundamentally misleading to attempt to identify factors related to achievement by relating them singly. The most common example of this dangerous approach is scouring simple 'one way' tables relating practice to achievement and identifying some factor as a significant explanation for achievement. The negative version of this approach is equally dangerous. For example, it is misleading to look at grade structure and say that this factor does not matter because some high achieving countries have seventh through ninth grades grouped while others have eighth through twelfth. To determine the importance of grade structure for US education, it must be placed in a US context. In this context, the grouping of eighth grade with lower grades combined with a repetitive curriculum, a lack of focus, and other factors, mutually determine the nature of the US eighth grade experience as typically more elementary than secondary.

Inherently, the above strategy makes no attempt to account for or assess the impact of the interaction (likely large and almost certainly negative) of explanatory factors considered one at a time. A focus on single factors may overestimate the importance of some factors that correlate with achievement but which have their impact mainly in interaction with other factors. Such a focus on single factors may also underestimate the importance of other factors that gain their impact primarily through interaction and thus do not offer by themselves strong relationships with achievement.

A special case of this mistaken strategy is concluding that a factor is not important in accounting for higher achievement because some higher achieving countries show one pattern for that factor and others a very different factor. What will be important to American education is, of course, the impact of that factor in the US context. The global search through single factors that might relate to achievement not only ignores interactions but explicitly ignores understanding relationships in the context of a particular educational system. Examining factors systemically, with interactions and in context, is not simple. Failing to do so is simplistic.

Thus, it seems logical to expect multi-factor systemic solutions and suspect single factor explanations. Sole factor explanations are logically possible but, both on the grounds of the nature of the problems and the empirical evidence, seem unlikely. However, it would also be misleading to overgeneralize and conclude that everything matters. Systemic problems are not pandemic (nor pandemonic), merely systemic. Not all factors in a system are implicated – or implicated equally – in any one problem. It would be a mistake to think so.

However, it seems to be a more common mistake to think that some one thing matters so much that it outweighs all other factors. It is just this error that is made when one seeks a panacea – a single solution that would solve all of our problems. In this case, it would mean a monolithic factor that universally distinguishes higher achieving countries from lower achieving countries regardless of content, population, and all other contextual factors that define the system of those top achieving countries. It would also necessarily need to be implementable in American education. Perhaps recalling wars against werewolves, vampires, and other agents of pandemonium – or more likely borrowing a metaphor already borrowed from medical research in dealing with literal pandemics – one current colloquial term for panaceas are 'magic bullets': solutions so powerful that they require no aid from any other factors. Simply put, we are saying that there are no 'magic bullets' to cure our educational ills and that it is a mistake to think so.

Why are we even talking about single-factor solutions? Historically, debate about US science and mathematics education has been rife with a number of initially hopeful, but inevitably doomed, proposed monolithic solutions. Unfortunately, this same search for 'magic bullets' has characterized much of the debate surrounding the TIMSS results thus far. It is a mistake the US can not afford. For that reason, we devote some space to examining some of the proposed 'magic bullets.'

THE QUEST FOR THE GRAIL:
WHERE ARE OUR MAGIC BULLETS?

Some of the earliest Western traditions of the search for panaceas for the world's ills were searches for holy relics, the prototype of which was the quest for the Holy Grail of Arthurian fame. Unlike literary Grail quests, the quest for solutions to US problems in mathematics and science education is not an end in itself nor is the journey its own reward. We may need our consciousnesses raised but, in the present context, what we most need are solutions. If there *is* a magic bullet, we need it and we need it now. Let's consider some of the candidates.

Assign more homework. Others have held that if US science and mathematics teachers would simply assign more homework this would be sufficient to produce desired achievements. We have. It hasn't. This perennial 'quick fix' may be valuable as part of a package of improvements, but there is empirical evidence which denies that it is sufficient alone to solve our problems.

We discussed earlier in this report that US mathematics and science teachers already assign more homework than is typical of their counterparts in many higher achieving TIMSS countries. If the suggestion here is that they assign yet more homework, then this implies one of two things: US students have inferior abilities in mathematics and science that can only be overcome by hard work and lots of it or; US students, with the aid of the teachers, have to become the sole solutions to systemic problems lying deep in the structural characteristics of US educational practice. The experience of various sub-national TIMSS replications (for example, Minnesota in science education) suggest that there is nothing inherently inferior in the abilities of US students (as does the comparative science achievement of US fourth graders). Further, if our educational problems are systemic as the TIMSS data suggest, no one or two components of our education systems are likely to be able to solve our problems by themselves. Teachers assigning and students doing more home-work seems a non-starter as a magic bullet to solve our educational ills.

Nonetheless, this proposal does contain an element of truth. US mathematics and science teaching uses homework in ways quite different from other countries.[1] Chapter 3 has shown, for example, that US science and mathematics teachers devote far more in-class time to allowing students to begin their home-work. Some help may be gained from exploring different approaches to the use of homework and to making it the out-of-class activity it was traditionally intended to be. However, such changes would have to be systemic. Individual teachers cannot easily create these changes. Students, as Chapter 4 discussed, have already come to expect in-class time to begin their homework. Success is likely to depend on consistent, widespread, long-term changes in this practice that persist in the face of initial resistance. Even then, however, it would merely contribute to solving our problems and would not be a panacea.

Get 'back to the basics.' For almost 30 years if not longer, American mathematics and science education have seen periodic cries for getting back to 'basics.' This typically includes a conception of subject matter and at least broadly concomitant approaches to teaching. The content is often arithmetic-oriented in contrast to attempted reform emphases on other content such as geometry or algebra.

The discussion of what we teach in Chapter 2 suggests that we typically have not strayed far from an arithmetic-centered mathematics curriculum through eighth grade. There is certainly evidence that other countries have included advanced content, a much richer kind of content 'core' or 'basics', and performed comparatively far better (as Chapter 5 showed). The case for mathematics seems clear. The case for science is less obvious but clearly suggestive. Certainly these data call into question not only whether getting 'back to the basics' is a panacea but also whether it is even a move in the right direction.

Devote more time to mathematics and science. Still others have held that the US could solve its problems if only it would devote more instructional time to mathematics and science education. As we saw in Chapter 2, the US does not differ appreciably from the other TIMSS countries in how much instructional time it allocates to science and mathematics. In fact, the US seems to be somewhat on the high side. Earlier studies showed that there was no clear or systematic pattern connecting overall instructional time in mathematics with national achievement.[2] Allocating more instructional time overall has always been a deceptive explanation of achievement. It is simply not a sufficient change by itself.

Perhaps the problem is in the definition of instructional time. The time actually devoted to mastering content is a product of many things: time spent on covering mathematics and science topics, time spent on new instruction rather than review, and time in which students are engaged in learning rather than simply attending class. Alternatively (or in addition), we could try to enhance the quality of that instructional time. This might take the form of better focused and less inclusive textbooks, or official curricula that did not call for time to be shared by so many different topics but rather focused it on a smaller number of strategic contents. If more time were allocated, it might permit not just additional time but increases in the quality of the time by including more challenging and demanding student tasks or by exploring the depth of topics rather than devoting time only to covering a broad range of topics or by doing both. However, all of the above considerations move the proposed solution from a monolithic single adjustment that would solve all problems to a set of related systemic changes in several factors. At this point, while time may be a focus of change, it has become a focus of systemic change that also requires concomitant changes in other factors.

Use of collaborative groups. Still others advocate the use of collaborative groups rather than whole-class instruction or individual seatwork as a kind of panacea for achievement concerns. Since little group work was reported in the TIMSS data, these data provide little evidence one way or the other about this proposition. On the other hand, the data showed the lack of group work in virtually all reporting countries.

Collaborative groups are often proposed as part of constructivist learning theories. The pedagogy mandated by such theories puts a premium on time for students to find their own conceptions that may be linked and fostered into something more like standard conceptions and process knowledge. That time can only come through increasing the amount of instructional time or through officially narrowing curricular focus to a smaller number of strategic topics so that teachers are free to develop those topics more fully and to omit others. Again, this has become systemic change rather than a monolithic solution.

Using technological advances more consistently in mathematics and science education. Others have been strong advocates for using technological advances more consistently in mathematics and science education. As a candidate magic bullet, this approach (as was that of collaborative groups) must be classified as 'untried' in terms of the TIMSS data. Those data showed US schools more often possessed appropriate technology and laboratory materials than many other countries. Many of the higher achieving countries both had less technology available and made less use of technology than did the US. The materials were more often available in US classrooms. However, they also showed a pattern of limited use either by policy or by characteristic practice. The case for the efficacy of using technology must remain at best 'not proven.'

Putting algebra earlier or perhaps pushing even more content down to earlier grades. Some have suggested that, based on what we have seen in other countries, many of mathematics' ills could be solved by putting important content earlier in our curriculum (for example, putting algebra in eighth grade or earlier). Similar proposals are made for various science contents. The most extreme form of both might be described as 'shoving all content down into earlier grades.' Unfortunately, the TIMSS data make it clear that the practice in countries that do treat such central contents as algebra and geometry earlier also includes organizing the content of even earlier grades in ways that differ significantly from US practice.[3]

Like many proposed monolithic solutions, this one has an element of truth. Placing in the middle grades significant and demanding content that currently is characteristic of US secondary school would at least break the pattern of considering seventh and eighth grades as the end of elementary school rather than the beginning of secondary. It should help to remedy the lack of rigorous, demanding content and to decrease content redundancy.

However, it could be that this proposal ignores the erosive and transforming power of US schooling traditions. Algebra is already an option for some US eighth graders. Lowered expectations, covering many simple topics, and re-packaging content in less demanding forms are well established US practices. It seems likely that if all eighth graders were to take algebra, algebra would become such that all eighth graders could take it. That is, the transforming effects of American educational practice would likely prove so strong that algebra would become something different from what it now is and, losing its demanding character and what intellectual coherence it possesses, would fail to be the solution for which we had hoped.

Perhaps the fatal evidence for placing content in lower grades as a 'magic bullet' is provided by the achievement results of countries that have done so. For example, take the case of algebra. Many TIMSS countries have distributed algebra's introduction and development differently in the grades and some have begun its study quite early. However, there is no systematic pattern in how early algebra is introduced and the level of achievement (even in algebra) of the countries making those introductions. It appears not to be simply a mat-ter of introducing this content early but of how it is introduced. Further, there is no guarantee that any one country's effective method would be transportable to American culture and schools.

As with many ideas, however, it would be a mistake to throw the 'baby' out with the 'bathwater.' Simply because introducing algebra, geometry, or any other content earlier is not a magic bullet does not mean that it could not be a component of solving some of our educational problems. Certainly, it might help remedy the lack of rigorous, demanding content in seventh and eighth grade curricula and might break the 'glass ceiling' that keeps those grades more commonly grouped with elementary grades. As part of the solution but not as magic bullet, proposals for introducing significant content earlier seem clearly worth discussing further.

Improve teaching and teacher training. Some have proposed that if we improve teaching and teacher training, all else will be well with US science and mathematics education. This might have potential as a cure-all if US teachers were the source or even the heart of our problems. As we discussed above and elsewhere, US mathematics and science teachers try to do the job we hand them. They are not responsible for unfocused, redundant, and unde-manding curricula. They are not responsible for cautious, inclusive textbooks.

Were we, in fact, to try to train teachers and improve their teaching suffi-ciently to overcome our current problems, we would have to begin to train them to do without their official curricula and do without their ubiquitous text-books. Improvement is not simply a matter of showing videotapes of Japanese mathematics teachers to US teachers and saying 'do it that way.'[4] Even better pedagogical approaches would not overcome discretized, highly redundant, and unchallenging content that persists in curricula. Focused, coherent, rigorous, and demanding curricula seem a necessary, but not sufficient, condition for improving the quality of teaching.

As with other proposed magic bullets, this one has grains of truth that should be carefully gathered and explored. We can improve teacher training. We likely can affect even more changes by serious changes in the working conditions and professional status of teachers. Doing either or both should contribute to solving our mathematics and science education problems. However, the problems remain systemic. Altering any single element of the system seems inherently unlikely to provide a complete solution. Expecting this to happen seems a clear case of confusing symptom and disease.

Centralize curriculum and educational decision making. Others advocate centralizing curriculum and educational decision making as a means to overcoming the inefficiencies and inequities in our approach to mathematics and science curricula and instruction. We are not among them. Certainly this might, in fact, be the root of a 'magic bullet' to significantly improve US science and mathematics achievement. It would surely strike at one of the structural features that characterize US educational practice.

However, the TIMSS data provide examples of countries with centralized curriculum and educational control that achieve both higher and lower than the US. Centralization in itself seems not likely to be a sufficient change. Even the suspicion of desires for partial centralization through national voluntary achievement testing have recently stirred the political climate so much that radical retrenchment or abandonment were needed.

It is not *centralization* but *coherence* that demands national and regional consensus. Unlike bureaucratic centralization, this consensual search for coherence would not inhibit innovation. Neither centralization by itself nor a national curriculum seems to offer a viable American solution to our problems. However, given our characteristic dispersed and shared responsibility for educational decisions, the fragmentation that results, and our market-driven textbooks, it is clear that some kind of coordinated effort is essential to any successful systemic solution. Its curricular vision is the core of any educational system. With a coherent vision, a coherent system is possible. With a fragmented vision, the system is likely to be more chaotic than coherent.

Somehow states and communities must work together to decrease fragmentation. We are not talking about a federal bureaucracy or even about national standards with some kind of federal sanction (or enforcement). There are other alternatives. One possibility would be voluntary regional alliances that get at something more specific than the policy level typical of the National Governor's Association summits and that seek to enhance and spread consensus. Finding this or some other kind of alternative to unacceptable centralization is not just for the greater aggregate ('national') good, but it also gives states and regions a more solid basis for dealing with a mobile population. Surely we must beware of going so far with devolution that community autonomy overcomes general welfare, even within states, and that healthy diversity gives way to dangerous fragmentation.

Turn more of schooling responsibility over to local and market-driven forces, site-based management of schools, multicultural education, magnet and charter schools, vouchers, and the running of public schools by private corporations. Is this a potential 'quick fix' for the *nation*? Some individual projects might succeed better than some current schools. However, to believe that this approach – which inherently increases fragmentation and lack of coherence – would raise our aggregate national levels of science and mathematics achievement requires a naive faith in market forces and the ability of separate small school units to define curricula that are 'world class.' Neither evidence nor logic support this position even though it, as do the other proposed 'magic bullets' may contain some grains of helpful truth.

Imitate the curricula or instructional practices of more successful countries. It is a surprise, and perhaps a sign of cultural naiveté, that some still propose that we can solve our achievement woes by imitating the curricula or instructional practices of more successful countries. This tangent in the search for 'fixes' occurs in both more and less naive forms. The more naive forms are obviously not feasible. However, those that offer such proposals are not always naive. A recent article suggested that TIMSS was fundamentally wrong in its data analyses and that one might better have studied what other countries do not like about their educational practices and areas for which they are trying innovations.[5] This article provides what appears to be thoughtful analyses, but the conclusion (studying innovations is better than studying what is unquestioned) seems to involve a more sophisticated form of the same logic of 'emulating other countries as a way to solve US problems' that it accuses TIMSS of using and that we are rejecting here.

Emulating seems to us to be inherently risky, whether we are emulating perceived strengths or innovations aimed at self-perceived weaknesses in other countries. Effective practices in all countries are linked to those countries' educational systems and cultures. The hope that one can find an innovation or established practice suitable for import as a solution to US policies would seem to be an example of the fallacy of 'deja do' – that is, 'we've already seen this work elsewhere so it should work for us.' That approach, in its more naive forms, betrays a fundamental misunderstanding of the use of cross-national studies. Certainly TIMSS is built around the idea that cross-national comparisons done carefully can enhance understanding. It is equally clear – and always has been – that emulating other countries practices is far from feasible, reasonable, or a magic bullet.

CONCLUDING REMARKS

In fact, the TIMSS data suggest that there is no one, monolithic factor that makes all the difference or that, if such a factor exists, it has certainly not yet been found among those most commonly suggested as quick fixes. The 'fixes' discussed in this chapter would be nothing more than 'straw men,' set up to be easily dismissed were it not that each has been seriously proposed by someone and most have elements of useful ideas that may help us better understand the systemic nature of the problems underlying US comparative science and mathematics achievements.

The US cannot afford to abandon the search for solutions to its difficulties. However, it can hardly further that search by looking for answers in all the wrong places. Quick fixes simply do not exist. They are the wrong place to look for workable solutions. We can examine them for insights that may help us recognize more adequate systemic solutions to systemic problems, but we cannot take them seriously as solutions by themselves. Mythical quick fixes offer no way to escape facing the consequences of our educational choices by pulling some quick fix 'rabbit' out of our collective hat. For US science and mathematics education, the quest is not the goal. Solving our problems is.

Notes–

[1] See Chapter 3 of this volume and also Schmidt et al. (1996). *Characterizing Pedagogical Flow.* Dordrecht: Kluwer.

[2] See McKnight et al. (1987). *The Underachieving Curriculum.* Champaign, IL: Stipes.

[3] See *Many Visions, Many Aims* (Vol. 1) for details on how various countries handle algebra and geometry. Schmidt, W. H., McKnight, C., Valverde, G. A., Houang, R. T., & Wiley, D. E. (1997). *Many Visions, Many Aims: A Cross-National Investigation of Curricular Intentions in School Mathematics.* Dordrecht: Kluwer.

[4] Certainly this would be an abuse of the research-oriented work of Stigler and his associates who have never advocated such an approach.

[5] Atkin and Black (Atkin, J. M., & Black, P. (1997). Policy perils of international comparisons: The TIMSS case. *Phi Delta Kappan.*, 22–28) end their article by writing, "We have tried to demonstrate that it is risky to come to policy conclusions on the basis of perceived relationship between student scores in a particular country and other aspects of that country's education system, specifically whether or not it has formal national standards for subject-matter content. For policy purposes, there may be greater value in examining what the various countries do not like about their current practices in mathematics and science education and are investing large

sums to change." We find it hard to disagree with the first part of the statement and, as this volume makes clear, we do not advocate examining only the relationship between student scores (whatever version of these the authors meant) and national subject-matter standards or any other single factor. The second part of the statement, advocating examining what countries do not like about their practices, may serve the purposes of Atkin and Black's involvement in a cross-national OECD study of innovations but seems flawed by the same logic they criticize. Why should emulating avoidance of undesired features of other countries' practices be any less risky than emulating features that appear to work? The more fundamental point is the risks of naive emulation.

Chapter 9
Some Stories TIMSS Can and Cannot Tell

As we begin to bring our discussions to a close, what can we say we have learned? Certainly, Chapter 8 was intended to bring home the point that we have not found any 'quick fixes' for US weaknesses in mathematics and science education. However, we did learn several important lessons from examining the TIMSS data, trying to see more clearly some key features of the current state of US science and mathematics educational practice, and trying to understand how those features of US practice may be linked to some of its deeper, more fundamental structural characteristics. More can be done and, in fact, we have progressed in our analyses in ways that have influenced our thinking about what to report here even though we cannot report all of those analyses yet since they remain to be completed. The authors believe that behind all of these analyses runs the story that what has happened in the US to education in mathematics and the sciences has been the consequence of our societal and educational choices.

Of course, there are a few stories the TIMSS data cannot tell. Every data collection, no matter how carefully conceived, developed, implemented, and organized has its limits. Some questions will always be beyond the scope of those that can be approached empirically through a data set due to those necessary limits. The first part of this chapter is devoted to summarizing some of the stories we believe the TIMSS data can tell. The second part focuses on the limitations of the TIMSS data and needed follow-up research.

A FEW STORIES FROM THE TIMSS DATA

Behind these data analyses and the underlying story that we set out as a hypothesis in Chapter 8, there are still a few more stories that the TIMSS data can tell. As part of our concluding remarks, we examine briefly seven of those cross-cutting themes here. We try to capture each of these stories or themes in a brief slogan and then explain the meaning behind it.

'Achievement is not achievement.' Cross-national comparative studies in education, especially the IEA studies, have always had at their center achievement testing in the participating countries. In most cases, cross-national comparisons were based on comparing global achievement scores. There have been a few exceptions. Comparative results on individual items have been a staple, at least of national reports based on participation in IEA studies. In some cases (for example, the Second International Mathematics Study) scores were reported only for separate categories and no global scores were constructed or, at least, reported.

TIMSS has been no exception. Thus far, global scores have been released for both mathematics and science and for both the pair of grades containing the majority of nine-year-olds in each participating country and the pair containing the most thirteen-year-olds. The global scores were scaled to form aggregate scores for 'mathematics' and for 'science.' Science and mathematics are within quotation marks in the previous sentence because the scaled scores presented – as is the case for any test scores – do not represent mathematics or the sciences or even school mathematics and science (those disciplines as they are presented in schools).

The scores represent school mathematics and science (and through them mathematics and the sciences) as they were captured through the particular items selected for the tests. While care was taken to obtain a representative sample of items covering major areas of school science and mathematics, as with every test they are a sample. In this case, the sample is constrained by limited testing time and the need for multinational consensus on the acceptability of each item. Some aspects of the two school subjects were better represented and others represented less well.

The reporting of a global, scaled score tends to hide this characteristic of achievement testing. Results for some specific items and for several reporting categories have been published to supplement the scaled total scores, but it is the latter that typically draws the most attention. This is important to note primarily because this kind of global total, especially when scaled to represent a level of attainment on some underlying trait (in this case, presumably, achievement or mastery of 'mathematics' and of 'science'), tends to focus only on what is in common to answering the set of items correctly. If that item set is disparate – as these are – the result is a measure of attainment on the general competencies needed to answer the variety of items correctly.

There is, perhaps, nothing inherently wrong with this approach although it seems easily used in unintentionally misleading ways. What *is* wrong, however, is taking results from these general scores that have enhanced commonalities and suppressed differences among disparate items and using those results as evidence that curricular differences are not relevant to explaining achievement or at least not as relevant as more general factors such as socioeconomic status.

Our analyses, especially in Chapter 5, suggest something quite to the contrary. When we construct achievement 'tests' of small sets of items sampling a specific area of school science or mathematics – and thus far less disparate than the full item set – the resulting scores seem much more clearly to vary and to vary in ways related to curricular emphases. Those findings have been discussed thoroughly in that chapter.

What we mean by saying that 'achievement is not achievement' is that the relevance of achievement results to school, curriculum, and instructional practices depends on how the achievement was measured and at what level of specificity. Specific achievement measures relate more strongly to school, curriculum, and instructional practices. There are measures that are more and less sensitive to educational factors. When using any achievement result, care must be taken to examine the specificity and nature of the achievement measure involved and how likely it is to be general (and thus relate more to general background characteristics) or specific (and thus relate more to curricular and school factors). This caveat is especially important in evaluating critically any claims for the irrelevance of curricular specifics to educational achievements.

'Curriculum matters.' Moving beyond global achievement measures, we can turn to the outcomes of instruction. A pressing question then becomes, 'Does it really matter if we emphasize a particular content in our curricula and in our teaching?' The answer from our analyses seems to be a resounding 'yes.' In a very important sense, what we teach is what we get and what is related to higher attainments for our students. There was considerable variation among achievement within specific content areas, both in mathematics and in science and at all grade levels tested. There was also considerable variation in the gains estimated and reported in Chapter 5.

Without trying to establish cause and effect, or even move to formal analysis of linkages, the patterns of higher and lower achievements seems on inspection to have been interestingly related to the patterns in curricular and teacher content emphases. Delivering emphases in curriculum and instruction might not be a 'quick fix' (especially in a context of competing demands for emphasis) but it was a factor that made a difference. In general, more emphasis was related to higher achievement, higher achievement gains, or both. This is what we are trying to capture with the phrase 'what we teach is what we get.'

'Some get and some don't get.' Mathematics and science education, as are most other subjects in US schools and classrooms, are delivered in an approach to education (and in the US' multiple education systems) that provides significantly different access to educational possibilities. Not all students have access to the same possibilities for learning. That differentiation, despite the intentions that may have led to it, is organized neither for efficacy or equity.

In Chapter 6, we suggested that differential access to educational possibilities was an important systemic characteristics of US mathematics and science educational practice. Evidence on what we teach to whom was presented in Chapter 2. Evidence on differing attainments in mathematics and science related to differences among schools, tracks, and classrooms was presented in Chapter 6. In Chapter 8, we summarized the main thread of these findings with the phrase that 'some get and some don't.'

The slogan is simple; the reality is not. It seems likely to represent a pervasive, fundamental characteristic of US educational practice in science and mathematics education. If it were necessary, we could do little other than to note its existence. If it were without consequence, it would not even be worth noting. Some differential access is inherent in our fundamental tradition of state and local responsibility for educational matters. Other differences are created by policies such as tracking.

These latter differences are not necessary; they are created by choice and are, therefore, far from being as fundamental as distributed responsibility for education. These latter differences likely have consequences. Some children achieve less, not because they work less hard or have less ability to master mathematics and science, but because of where they attend schools and the policies that determine which educational possibilities they will have access to. To not be able to be the best that we can be is sad. To have this true by policy is ethically bankrupt, if not socially criminal. Some get and some don't get and we have chosen, perhaps without knowing, to make this true through our practices of differential access. That very characteristic American practice is very un-American.

'It's not just how long you make it but how you make it long.' It's sad to find one of the better ways to capture a fundamental fact in a slogan so similar to one that once served in a cigarette advertisement. Unfortunately, this is one of the pithy ways to capture two kinds of findings discussed in Chapters 2 and 4. First, how long we spend on various topics in mathematics and science *does* matter, but achievement is not just a matter of that. The quality of the instructional time devoted to a topic, how we 'make it long', also matters.

Fragmented instructional time, especially as divided among 'microbursts' of content and short-term classroom activities, produces not only quantitative differences in instructional time for different contents but qualitative differences as well. Balancing demands for on-the-spot review, new instruction, and in-class time for 'homework' to secure student engagement creates almost impossible instructional time management challenges. US mathematics and science teachers are masters at that kind of management. The question is, 'Should they have to be?'

Behind our slogan on the importance of qualitative as well as quantitative differences in instructional time lie real questions about directions for needed change. Do we truly need as much on-the-spot review as we characteristically provide? Are there alternatives to using in-class time for students to begin 'homework' and more in-class time to evaluate that homework, or are there alternatives that could be made to work in US mathematics and science classrooms that would make more time available for instruction on new content? Are there ways to transform US science and mathematics teaching from orchestrating microbursts of five, six, and more activities to sustaining a nuanced development of a single integrated piece of new content with some depth and more challenge for our students? If we can get to 'yes' on any of these questions, we will have important information that could lead to greater educational attainments in science and mathematics.

'Pedagogy is conditional.' This slogan is more opaque than many of the others. What aspects of pedagogy? Conditional on what? By this slogan we mean to imply that the use and effectiveness of pedagogical approaches are situated inextricably in culture contexts in complex ways and thus cannot be treated as unconditioned strategies equally at home in different cultural contexts. It is this that lies behind why emulating the practices of successful countries simplistically not only offers no 'quick fixes', but likely no solutions at all.

Pedagogy is complex, as we saw in Chapters 3 and 4 when we examine the beliefs and pedagogical choices of US mathematics and science teachers. Pedagogical choices are situated, even in the same classroom. They depend on specific contents. They depend on specific goals and tasks to be accomplished (introduce new content, clear up a pervasive misconception, and so on). If pedagogy is so situated and complex across situations in a single classroom, how could it be otherwise across cultural contexts in different countries?

It is these ideas that we try to capture by saying 'pedagogy is conditional.' We must learn the limits of what we can learn from cross-national comparisons. They will not be the sources of 'quick fixes.' They will not even be the source of things to emulate without careful study of their ties to cultural contexts and how well these accord with or can be adapted to the realities of US classrooms. We must also learn that pedagogy is conditional and complex even in US classrooms. Simple policies will not make US mathematics and science pedagogy simple. Changes may be influenced by policy; they cannot be dictated by it.

'Mobility matters.' School officials and teachers perceive themselves as having to deal with a mobile, changing student population. The perception seems to be grounded in reality. For example, in the classrooms sampled as a part of TIMSS, only 83 percent of those who began the year in a class ended it there. Further, about 15 percent of the students in each class (on average) at the end of the year, came into the class sometime during the year rather than beginning the year in that class. It seems likely that some of these moves were among classes within the same schools. Even so, a large proportion of teachers really do end the year with different students than those with which they began the year. A significant proportion of students really do end the year in different classrooms than those in which they began it.

The US national context is one of varied educational possibilities. The specific science and mathematics curricular possibilities to which a student will have access, even in the same grade, even in the same state, and sometimes even in the same community or school, will vary depending on the specific state, community, district, school, or class in which the student happens to be. Given that context and given real mobility among American families, many students will be exposed to varied education systems within the US, a land of many and varied systems.

Exposure to different systems and the differing access to educational possibilities that results affects the mathematics and science attainments of those students. It also affects what teachers must assume about their classes and thus which strategies they consider appropriate (for example, the level of review in a given grade). So, in a significant and pervasive sense that affects both the substance of instruction and its outcomes, mobility matters.

'Connections count.' US students, living in an MTV world and a world of both channel and World-Wide Web surfing, require considerable stimulus to engage and hold their attention. Unfortunately, only a small portion of higher mathematics and science achievement seems particularly linked to areas that interest students and that connect to their everyday experiences and concerns (e.g., the effectiveness of ecology-oriented questions in the US).

Today, more than ever, curricular content can ill afford to be without connections to those experiences and concerns. Instruction can ill afford to capitalize on those connections, even as it develops basic content. Artificially engaging 'learning' environments created by boosting the levels of stimuli through constant transition among brief activities and short subject matter tasks, may maintain superficial engagement for a time. As we saw in Chapter 4, however, eventually this boosting must loose its impact and students realize the content to which they are exposed for how superficial it is since, at some point, they begin to find it easy, largely boring, and are willing to admit that their peers, if not themselves, find it unimportant.

The only answer to this would seem to be to demonstrate its importance over and over, not in artificial ways but by linking it to real concerns of students. Perhaps this also implies that students might be more genuinely engaged by longer, less routine learning tasks in which different phases of a complex task connect and unfold over time. In either case, 'connections count.'

SOME STORIES TIMSS CANNOT TELL

The TIMSS data are complex. Their analyses are also complex — if done carefully, keeping each piece of data in context. The result of these data and their analyses are many, many stories. We have tried to build this volume around themes in those stories. However, there are some cross-cutting themes that seemed to emerge from surveying the main story threads of this volume and the array of analyses thus far accomplished. The preceding section attempted to point out a few of those cross-cutting stories.

Good research admits the limits of its data. That does not mean that there is nothing that can be said about what might be done to go profitably beyond those limits in other research. This section examines some limits of the TIMSS data and examines some needed follow-up research.

The TIMSS data resulted from an exceptionally careful development and collection process. The entire process was model-centered and model-driven. That doesn't imply that every person currently involved with TIMSS subscribes fully to the model, but even those doing limited tasks and essentially unaware of that central model work with instruments and data that were shaped by it. A separate multinational project was responsible for development of data collection instruments and procedures for analyzing textbooks and curriculum documents. That project team worked on those instruments using preliminary data from classroom observations, teacher logs, and teacher interviews. The prototype instruments were then turned over to more inclusive, multinational committees for further development. They were reviewed several times by all participating countries and underwent both an early limited pilot test and a later more extensive field test in all countries participating at that time.

The achievement tests went through similarly careful design. Banks of items – short answer, multiple choice, extended response, and performance assessment items – were developed by teams, committees, and individuals from several countries. Scoring rubrics for free response items were developed by a separate multinational committee. The test design was developed and scrutinized by a multinational technical advisory committee. Items were selected to fit content coverage blueprints based on the test design and on early results from analyzing curriculum documents. This selection was largely carried out by a subject matter advisory committee made up of mathematics and science education experts from several countries. All items were piloted and the results screened by the technical advisory committee. This complex development process continued until the tests took final form. Sample designs and their implementations were refereed by a committee of sampling specialists.

Each country collected its own data which were forwarded to various international sites for cleaning, weighting, scaling, and, finally, analysis. Most countries analyzed their own data. Various TIMSS centers produced reports and monographs reporting achievement results, the project model, the curriculum frameworks used, results of curriculum and textbook analyses, and preliminary results from the various TIMSS questionnaires (especially those for teachers and students). That process continues.

The TIMSS US Research Center was responsible for analyzing official curriculum documents, textbooks, and other curriculum data for all countries, and for preparing reports on those data. They were also responsible for integrative reports on US participation. The US National Center for Education Statistics issued reports on achievement results that reported global results as well as a sampling of data from other instruments. The current report is the first US report that presents more detailed achievement test analyses, more extensive teacher data, and attempts to integrate these into an interrelated whole in which each part can shed light on the others.

In spite of this careful development, analysis, and reporting, there are still holes, limits on what the TIMSS data can 'say' empirically. This volume has suggested that the data thus far have given up several insights that may help the move towards more effective US mathematics and science education policies. Apart from policy considerations, the analyses also seem to stand as a revealing research portrait of US science and mathematics educational practices and their relationships with the nation's many education systems and with structural features of its general educational practice.

Not all of the TIMSS results have been reported nor have all of the data been released. Primary analysis is still going on following the agenda formed around the guiding model of educational opportunity or, alternatively, access to educational possibilities. However, even now, some of the limits of the TIMSS data have become clear and with that clarity has come some early insights for further data collections that could build on TIMSS in fruitful ways. Further primary analyses and the beginnings of secondary analysis, through which fresh eyes re-examine the TIMSS data, will reveal both further limits and further insights into needed research. Here we try to summarize only a few emerging insights.

The TIMSS data are of three types: achievement test results, document analyses, and responses to survey questionnaires. Each of these three types occurred in several complex forms (performance assessment as well as multiple-choice achievement items, partitioning and coding of entire textbooks, questionnaires not only to school officials, teachers and students but also to country representatives who were led to provide data for a complete portrait of the structure of their country's education system). In the end, however, the TIMSS data are still test results, analyzed documents, and survey responses.

Each of these data types has its limits. The strengths and weaknesses of achievement testing have been staples of the research literature for years and a number of specific limits for TIMSS have been discussed in this report. Document analyses are less familiar and the techniques more novel but for TIMSS they involved national teams following standardized procedures after face-to-face training to segment textbooks and official curricula documents, coding them into a set of subject matter categories and student performance task categories, and recording this coded information for central analysis without translating entire documents into a common language.

Quality assurance achieved high reliability for these analyses. However, they are still limited by the kinds of things they chose to consider (kinds of 'building blocks' for textbook 'lessons', etc.) and the categories into which subject matter and student performance expectations were sorted. A finer grained category set might yield a more refined and illuminating picture but could not be applied to the full set of documents since the original category set was applied by native speakers and the documents remain collected but untranslated.

The survey instruments, no matter how careful their design, could not ask all relevant questions. Indeed, several survey questions in the original instrument design were deleted before data collection in order to meet essential requirements that limit the amount of time required to complete an instrument. These survey instruments also have all the familiar weaknesses of self-report data.

Each of these data types have procedural weaknesses that limit validity and require 'triangulation' among the various data sources to assess validity and to build more complete analyses. Beyond these matters, the value of the data are limited by what is actually selected to be analyzed. It will be some time before the limits of the data are carefully assessed and before they have been made to tell all the stories that they can tell.

After all this, what stories will remain untold? What can we *not* learn from the interrelated, complex TIMSS data that still flows from achievement tests, survey responses, and the analyses of formal documents? First, all these data essentially are quantitative. They deal with what can be codified, measured, and counted. They cannot substitute for research that gets at words and ideas.

For example, how do teachers in different countries conceive of homework? The TIMSS data can tell us something about what teachers assign as homework, about how often they assign homework and how often they check the results in class. The data can tell us how much time students report they spend on homework and how many of them find it easy. These things can be quantified and asked in questionnaires. The TIMSS data, through analyzing textbooks, can even tell us about the specific content and what kinds of things students are expected to do with that content in the problems in their textbooks that are available for teachers to assign as homework. All of these things can be measured or classified and counted.

The TIMSS data cannot tell us which problems were actually assigned and thus what was actually expected from students. Do Japanese teachers focus on process rather than answers in homework assignments as has been widely reported? Do French teachers require students to review their class notebooks and to do problems given in class that are not in textbooks? Do teachers in all countries view homework as a means to an end and do they view this 'means' in the same way? These things require observation, interviewing, and listening – communication through words rather than counting and measuring. TIMSS data can provide quantitative information. The questions here require 'qualitative information' – information on qualitative differences among teachers and classes. The TIMSS data can provide a context for answering these questions. For example, knowing how often and how much homework is characteristically assigned by teachers in a particular country offers clues to how they may conceive of homework and its uses but cannot directly reveal those conceptions. Of course, homework is just one example of a source for questions that require qualitative data beyond the scope of TIMSS.

Overall, the TIMSS data can provide context for understanding that can be gained by qualitative research methods but they cannot substitute for such methods. They can target questions and phenomena in need of better understanding but they cannot provide that understanding. At the very least, follow-up case studies, observations, artifact collections (videotapes, actual student papers, recorded interviews, etc.) and analyses, and other qualitative research will be needed. Still many question remain to be investigated. The TIMSS data are full of anomalous successes, for example, unexpectedly high performing schools. Case studies of these anomalies might be done after the time- and resource-intensive quantitative research needs have been targeted on well-chosen, potentially fruitful cases by examination of the TIMSS data.

Second, some questions approachable by quantitative data cannot be answered by the TIMSS quantitative data. For instance, we have pointed out practices we believe to be fundamental structural and systemic characteristics of US education generally and in mathematics and science education in particular. For example, we discussed decentralized decision-making and educational responsibility. We believe these characteristics relate to features that we examined empirically through the TIMSS data. However, there is a need for a direct investigation of these systemic characteristics and their consequences. In particular, our analyses do not establish that changing these structural characteristics will improve science and mathematics education in this country. They also reveal nothing about how these characteristics may be changed or the impact of the changes.

Third, several times in our analyses we noted concomitant factors and correlates in our discussions but the interrelationships remained inextricably tangled. For example, the relationships between the lack of articulation in US education systems in different locales, perceived mobility of students, teacher on-the-spot review during instruction, and redundancies in official curricula seem clearly to have been linked. Can cause and effect be assigned to the elements of this tangle, or are these factors so inextricably interwoven that their interrelations cannot be further understood? Though the data we have can shed no further light – at least with the present analyses – surely these are factors that can be examined further, both separately and together.

The basis for the common US practice of in-lesson review, for example, can be explored through observing such review more closely, through teacher and student interviews, through examining the beliefs of school officials and those who shape state and local curricula. Such an emphasis on review seems not to have characterized US mathematics and science education in earlier days. Whether this is true can be verified through tracing the history of in-lesson review, how it changed, and what has made for the changes. Within the multitude of US education systems, schools, and classrooms, some seem to provide instruction characterized by more review than others. Certainly, teachers differ in the extent of their use of in-class review. Surely these differences would allow smaller-scale comparative studies, using qualitative methods, quantitative methods, or both, to get at the why's of the differences. There are also differences among countries. Many other countries do not seem to have in-class review as a characteristic practice. Further studies could be targeted on these countries' practices to try to separate cultural factors from those endemic to mathematics and science education anywhere.

Other tangles were noted. For instance, it seems likely that there were interrelationships between student attitudes (boredom with mathematics and science classes, thinking them easy) and teachers' use of in-class time for homework. We have speculated in the full book that this may be considered necessary to secure student engagement sufficient to make more likely the completion of homework tasks out of class. This relationship can surely be studied further, even were assignment of cause and effect never able to be established. US teachers differ in their use of in-class time for homework, although this is a broadly characteristic feature of US science and mathematics instruction. The differences among teachers could be studied, however, both to explain them and to shed light on the relationships between certain student attitudes and teacher practices. Certainly the practice of doing homework in class was not common in many other countries.[1] More targeted, smaller scale cross-national studies could shed light on these relationships.

There are many research stories important to US science and mathematics education and to policies likely to increase their effectiveness. TIMSS has been able to tell a surprising number of these stories, perhaps because of its scale and complexity. It promises to tell even more. However, after all is said and done, more will remain to be said and done. We are not in the business of research-as-usual, although we must carefully conduct the research in which we do engage. Surely all of us are joined in the business of understanding American education and facilitating its improvement. There are stories that TIMSS could not tell and never will be able to tell. Some of these stories need to be told. Careful examination of TIMSS reports and analyses can help identify at least some of these stories yet to be told. Studying the limitations of the TIMSS data may help target and guide new studies to tell more of these stories.

Notes-

[1]   In the development work planning the TIMSS instrumentation, we found that homework was a concept whose meaning varied among different countries (See Schmidt et al., (1996). *Characterizing Pedagogical Flow*. Dordrecht/Boston/London: Kluwer.). Following up on this was beyond the main scope of the developmental project and the project's resources. However, this seems an important feature to understand better, perhaps through multinational case studies. If such understanding resulted, it might provide highly relevant information on the necessity for and factors influencing US homework practices.

Chapter 10
What's the Next Story?

The TIMSS story at times seems to be never-ending. It is not; TIMSS will end. Eventually the time, resources, and energy for primary analyses will run out and the stories that follow from those primary analyses will trickle to a halt. However, that time is not yet. There are more stories that TIMSS will tell and some of them will be told soon. This brief coda is meant simply to point to one next phase in TIMSS analyses.

As this report neared final preparation, the initial report on the TIMSS Population 3 (final year of secondary school) was released.[1] It includes data (primarily on achievement) both for specialists that had studied science, mathematics, or both throughout their school years, as well as for a representative sample of all students in their last year of secondary education. Specialists were defined in TIMSS as those who had studied advanced mathematics (mathematics specialists), physics (science specialists), or both. They were tested on whichever of these two subjects they studied, or on both if they studied both. The representative sample of all students at the end of secondary school (which included the specialists) was tested for quantitative and scientific literacy considered appropriate, by consensus among TIMSS countries, for an educated citizen.

While detailed results can be found in that volume, a summary of some major findings may be useful here. These include:

MATHEMATICS AND SCIENCE LITERACY

- The Netherlands and Sweden were the highest achieving countries on the literacy test for the more general population. This was true for mathematics literacy as well as for science literacy (although the order of the two countries were reversed). The United States performed below the international average.

- How selective national education systems were or the methods used for selecting the samples tested had little effect in the literacy testing.

For example, the same two countries had the best performances when examining just the top 25 percent of the students in each country.

- In most countries there was a significant gender difference in mathematics and science literacy with males somewhat outperforming females. This was not true in the US for mathematics literacy but was for science literacy.

- Students who were no longer taking mathematics in their final year of secondary school did less well on the mathematics literacy test than those who were. A similar finding held for science literacy in most countries. The proportions of students still taking mathematics in their final year varied considerably from about 85 percent in nine countries to only about two-thirds of the students in several countries (including the US). Taking mathematics in the final year includes, but is not limited to, those taking advanced mathematics. Even smaller proportions typically were taking science in their final year.

- In most countries, more than 80 percent of the students tested reported using calculators at least weekly (in school or out). Students were given the option of using calculators on this set of TIMSS tests and most students reported making moderate calculator use on the test. Frequency of calculator use was related positively to both mathematics and science literacy in all countries.

ADVANCED MATHEMATICS

- France outperformed other countries for the advanced mathematics tests, although top performing countries from previous populations' TIMSS tests (e.g., Korea, Japan, Hong, Kong) did not take part in this test. The United States ranked next to last in advanced mathematics, performing well below the international average. Examining smaller proportions of high achieving students (the top five and 10 percent) showed the US to maintain its comparatively low ranking.

- There were significant gender differences favoring males in all but five countries. The US showed this gender difference.

- Most countries showed differing strengths and weakness for the three specific topic areas in advanced mathematics. The US ranked next to the bottom for the 'numbers and equations' area, at the bottom for the calculus area, and far below all other countries in the geometry area.

- Calculator use was reported as very common for advanced mathematics students in their final year of secondary school. Most reported moderate calculator use on the TIMSS test and more use was positively related to achievement in most, but not all, countries.

- The United States was one of only three countries taking both the mathematics literacy test and the advanced mathematics test that scored below average on both tests.

PHYSICS (ADVANCED SCIENCE)

- Norway and Sweden significantly outperformed the other countries in physics (advanced science), although top performing countries from previous populations' TIMSS tests (e.g., Korea, Japan, Hong Kong) did not take part in this test. The United States ranked last. The US ranking was still last when comparisons were restricted to the top five and 10 percent of the students taking physics.

- There was a significant gender difference favoring males taking the physics test in the United States and all other participating countries but one. The size of gender differences varied for specific content areas within physics. Roughly equal proportions of males and females took physics in their final year in the US and four other countries but certainly not in all participating countries.

- Countries varied in their relative standing for the five different content areas that comprised the physics test. The Unites States scored below the international average in all five areas. In two areas the US ranked next to the bottom, in one other area they tied for the bottom ranking, and in a fourth area they achieved sole possession of the lowest ranking.

- As in advanced mathematics, most physics students reported frequent calculator use and most reported using calculators on at least some questions of the TIMSS physics test. A positive relationship was found in most countries between general calculator use and achievement on the physics test, although the relationship was not as strong as that for mathematics. How much calculators were used on the test itself was not consistently related to achievement in all countries, although students who reported not using calculators at all consistently scored below those who reported some calculator use.

These findings are important and informative. These data also offer several possibilities when further analyzed and related to other TIMSS data. The US (and other countries) may learn just what science and mathematics end-of-secondary students take away from their years in school, and just how they compare with their counterparts elsewhere. More detailed evaluation of US twelfth-grade physics and calculus course contents should reveal important comparisons with those contents in other countries and the linkage between course contents and achievement.

Those answers will be interesting and important. These answers and further analyses will also help to answer questions already raised by analyses of US mathematics and science curricula, textbooks, and teaching. They will complete the sequence of comparisons that began with the release of results for seventh and eighth graders and for third and fourth graders.

We raised some of those questions in this report. US curricula through eighth grade are 'a mile wide and an inch deep.' They appear to proceed by assuming that learning small pieces of content will lead to cumulative mastery at some point when all of those pieces are brought together. This pattern certainly seems true for mathematics. To the extent that science emphasizes factual knowledge, it is likely true for science as well, as students gradually assimilate the body of traditional scientific facts and processes and to meet goals that seem, at times, to imply more 'knowing about' science than 'knowing' science.

Our analyses in this volume and those in *A Splintered Vision* suggest that after eighth grade, specialization sets in mathematics curricula, textbooks, and courses. US high school grades are where it presumably 'all comes together' for US science and mathematics education. Unfortunately, only about 20 percent of US students are in algebra classes in eighth grade and even fewer will continue mathematics study. Similar patterns hold for science study. This suggests that those who do take calculus and physics will be specialists indeed but only a small percentage of those who finish high school. Still others may take some high school mathematics and science but discontinue those studies at some point in their high school years.

Can a few years of specialized, somewhat more focused study after years of incremental and cumulative mastery produce levels of mastery by the senior year that are comparable to or even superior to those of countries who have taken a more focused and less incremental approach to science and mathematics education across many grades? Will it, in fact, ever 'all come together' for US students or will the accumulated increments never coalesce into significant cumulative mastery? If it does not, this must surely call into question this whole approach to structuring mathematics and science curricula.

The achievement results for US fourth and eighth grade students did not bode well for the relative standing from these twelfth grade results for specialists. Now completed, the trend was clearly one of decline in relative standing. In science, we moved from among the best among TIMSS countries at fourth grade to being slightly above the cross-national average at eighth grade to a markedly lower standing in physics. In mathematics, we moved from being slightly above the cross-national average at fourth grade to being essentially at the average (or below it in some areas) at eighth grade to being clearly below average in advanced mathematics and mathematics literacy. Preliminary analyses have begun but more detailed answers wait for further work.[2]

Our population of general end-of-secondary students did not do well, either in mathematics or science literacy. They had either taken less demanding courses or discontinued the study of science and mathematics at some point (after its mandatory study in eighth grade but before twelfth grade). *A priori*, it was far less likely to 'come together' for them since they likely had not taken all or, at least, the most demanding of mathematics and science 'capstone' courses offered in US high schools. Their accumulated science and mathematics mastery seems clearly to have been insufficient to make them comparably literate to their counterparts in other TIMSS countries (although detailed results await further analyses).

Is the mathematics and science pursued in US curricula even that which is most appropriate for scientific and quantitative literacy? Is there to be a conflict between preparing specialists and preparing citizens or can we do both? Further analyses of our twelfth grade results should contribute to the answer to those questions.

These results are important. They will give us a more complete picture of the 'yield' of school science and mathematics and our efforts at educating US children in mathematics and sciences, a picture that may help us better estimate the urgency of change or the lack of that urgency. Based on what we have seen in the analyses so far, we are not expectant.

Finally, these results face us with the urgency of national weaknesses in mathematics and science education. US public debate about the state of American mathematics and science education has already used these results to support previously held positions. Thus far, since the analyses are only beginning, the TIMSS data have been used, at best, illustratively. Their true significance can only be revealed by further careful empirical investigation, analyses which have yet to be completed or reported. Until the time when these results do become available, care must be taken in evaluating claims for 'solutions' that use merely the broadest strokes of the TIMSS data reported to sound the cry that there is a problem. Perhaps when the careful analyses are in, Americans will more rationally face the consequences of their educational beliefs and choices and become serious about finding solutions to the deep, systemic problems that remain intractable to piece-meal solutions.

US citizens want American education, education built around our deepest beliefs. We want education that leaves America as America, pre-eminent, a land of hopes and dreams. We want to deny that we don't have those things now. If, as a nation and as individuals, we deeply want both, we can no longer afford to look away from our problems, nor can we accept proposals that merely attempt to 'use' TIMSS findings for their shock value rather than their empirical insight. It seems likely that the more careful analyses soon to emerge will suggest that we must face the consequences of America's national choices and how we must face them, understand them, and together seek new solutions.

Perhaps then we may finally put aside private agendas and the 'education wars' to which they have led to pursue a public agenda of fundamental systemic change in science and mathematics education. Anything less does not accord with our own values. Anything less will lead to the death of hopes and dreams – not just ours, but those of our children, America's future.

Notes–

[1]   Mullis, I.V.S., Martin, M.O., Beaton, A.E., Gonzalez, E.J., Kelly, D.L., and Smith, T.A. (1998). *Mathematics and Science Achievement in the Final Year of Secondary School: IEA's Third International Mathematics and Science Study.* Chestnut Hill, MA: Center for the Study of Testing, Evaluation, and Educational Policy, Boston College.

[2]   See McKnight, C.C., and Valverde, G.A. (in press). In G. Kaiser, E. Luna, & I. Huntley (Eds.), *International Comparisons in Mathematics: The State of the Art.* London: Falmer Press.

Appendix A–Table 1. Descriptions of TIMSS Student Populations in Each Country Included in this Book.

Population 1	Lower Grade			Upper Grade		
	[1]Years of Formal Schooling	Mean Age	Mean % Repeating Grade	[1]Years of Formal Schooling	Mean Age	Mean % Repeating Grade
Australia	3 or 4	9.2 (0.0)	0.4 (0.1)	4 or 5	10.3 (0.0)	0.6 (0.2)
Canada	3	9.1 (0.0)	1.3 (0.2)	4	10.0 (0.0)	1.1 (0.2)
Czech Republic	3	9.4 (0.0)	1.0 (0.2)	4	10.4 (0.0)	1.2 (0.2)
England	4	9.1 (0.0)	0.1 (0.0)	5	10.0 (0.0)	0.0 (0.0)
Hong Kong	3	9.1 (0.0)	1.5 (0.2)	4	10.1 (0.0)	1.8 (0.2)
Hungary	3	9.4 (0.0)	*	4	10.4 (0.0)	*
Israel				4	10.1 (0.0)	7.3 (4.7)
Japan	3	9.4 (0.0)	0.0 (0.0)	4	10.4 (0.0)	0.0 (0.0)
Korea	3	9.3 (0.0)	0.1 (0.1)	4	10.3 (0.0)	0.2 (0.2)
Netherlands	3	9.3 (0.0)	2.8 (0.6)	4	10.3 (0.0)	2.3 (1.5)
New Zealand	3.5-4.5	9.0 (0.0)	0.2 (0.0)	4.5-5.5	10.0 (0.0)	0.3 (0.2)
Norway	2	8.8 (0.0)	*	3	9.9 (0.0)	*
Singapore	3	9.3 (0.0)	0.0 (0.0)	4	10.3 (0.0)	0.0 (0.0)
Thailand	3	9.7 (0.0)	1.6 (0.4)	4	10.5 (0.0)	1.9 (0.5)
United States	3	9.2 (0.0)	1.0 (0.2)	4	10.2 (0.0)	0.8 (0.2)

Population 2						
Australia	7 or 8	13.3 (0.0)	0.1 (0.0)	8 or 9	14.2 (0.0)	0.1 (0.0)
Belgium(Fl)	7	13.1 (0.0)	2.2 (0.3)	8	14.1 (0.0)	4.2 (0.5)
Canada	7	13.1 (0.0)	2.2 (0.3)	8	14.1 (0.0)	2.8 (0.6)
Czech Republic	7	13.4 (0.0)	1.8 (0.2)	8	14.4 (0.0)	0.5 (0.1)
England	8	13.1 (0.0)	0.0 (0.0)	9	14.0 (0.0)	0.1 (0.0)
France	7	13.3 (0.0)	12.4 (0.6)	8	14.3 (0.0)	9.0 (0.9)
Germany	7	13.8 (0.0)	5.8 (0.9)	8	14.8 (0.0)	6.3 (1.0)
Hong Kong	7	13.2 (0.0)	2.9 (0.5)	8	14.2 (0.0)	2.4 (0.3)
Hungary	7	13.4 (0.0)	*	8	14.3 (0.0)	*
Israel				8	14.1 (0.0)	0.1 (0.0)
Japan	7	13.4 (0.0)	0.0 (0.0)	8	14.4 (0.0)	0.0 (0.0)
Korea	7	13.2 (0.0)	0.1 (0.1)	8	14.2 (0.0)	0.0 (0.0)
Netherlands	7	13.2 (0.0)	1.7 (0.3)	8	14.4 (0.0)	4.6 (0.8)
New Zealand	7.5-8.5	13.0 (0.0)	0.1 (0.1)	8.5-9.5	14.0 (0.0)	0.1 (0.0)
Norway	6	12.9 (0.0)	*	7	13.9 (0.0)	*
Russian Federation	6 or 7	13.0 (0.0)	1.9 (0.2)	7 or 8	14.0 (0.0)	2.1 (0.3)
Singapore	7	13.3 (0.0)	0.1 (0.0)	8	14.5 (0.0)	1.1 (0.1)
Spain	7	13.2 (0.0)	8.8 (0.7)	8	14.3 (0.0)	12.7 (1.0)
Switzerland	6 or 7	13.1 (0.0)	1.2 (0.2)	7 or 8	14.2 (0.0)	2.8 (0.4)
Sweden	6	12.9 (0.0)	0.2 (0.1)	7	13.9 (0.0)	0.0 (0.0)
Thailand	7	13.5 (0.0)	0.0 (0.0)	8	14.3 (0.0)	0.3 (0.3)
United States	7	13.2 (0.0)	1.7 (0.2)	8	14.2 (0.0)	1.4 (0.3)

[1]   Based on the International Standard Classification. Within country differences are noted if areas or regions within the country have differing policies regarding when children begin school. In New Zealand, the number of years of schooling may vary since students begin schooling on or near their 5th birthday.

*   This question not asked in these countries.